THE METHOD OF
MATHEMATICAL
INDUCTION

INDUCTION IN
GEOMETRY

THE METHOD OF
MATHEMATICAL
INDUCTION

I. S. Sominskii

Translated and adapted from the fifth Russian edition (1959)
by Luise Lange and Edgar E. Enochs

INDUCTION IN
GEOMETRY

L. I. Golovina and I. M. Yaglom

Translated and adapted from the second Russian edition (1961)
by A. W. Goodman and Olga A. Titelbaum

Dover Publications, Inc.
Mineola, New York

This volume includes both *The Method of Mathematical Induction* and *Induction in Geometry*. Both books can serve as an introduction to mathematical induction for solving analytical problems, and they contain numerous examples from the fields of algebra, geometry, and trigonometry. Many of the problems are solved in the text, while others are left for the reader to determine.

Bibliographical Note

This Dover edition, first published in 2019, is an unabridged republication in one volume of *The Method of Mathematical Induction* and *Induction in Geometry*, both originally published in 1963 by D. C. Heath and Company, Boston, as part of the Survey of Recent East European Mathematical Literature series.

Library of Congress Cataloging-in-Publication Data

Names: Sominskiæi, I. S. (Il§ëìia Samuilovich), author. | Golovina, L. I. (Lidiëìia Ivanovna), author. | ëIiAglom, I. M. (Isaak Moiseevich), 1921-1988, author.
Title: The method of mathematical induction / I.S. Sominskii. Induction in geometry / L.I. Golovina and I.M. Yaglom.
Other titles: Metod matematicheskoæi indukëtìsii. English | Induction in geometry
Description: Dover edition. | Mineola, New York : Dover Publications, Inc., [2019] | An unabridged republication in one volume of: The method of mathematical induction, and Induction in geometry, both originally published in Boston by D.C. Heath and Company, 1963, as part of the Survey of recent East European mathematical literature series.
Identifiers: LCCN 2019018298| ISBN 9780486838564 | ISBN 0486838560
Subjects: LCSH: Induction (Mathematics) | Geometry—Problems, exercises, etc. | Logic.
Classification: LCC QA9 .S5813 2019 | DDC 511.3/—1dc23
LC record available at https://lccn.loc.gov/2019018298

Manufactured in the United States by LSC Communications
83856001
www.doverpublications.com

2 4 6 8 10 9 7 5 3 1

2019

THE METHOD OF
MATHEMATICAL
INDUCTION

I. S. Sominskii

PREFACE TO THE AMERICAN EDITION

THE METHOD of mathematical induction is a widely used method of mathematical proof. Without command of this method it is impossible to make a serious study of mathematics. Moreover, the basic idea which underlies the method of mathematical induction is interesting not only to students of mathematics and the applied sciences, but also to students in other fields.

The fundamental ideas of the method of mathematical induction, and some of its simpler applications, are presented in Chapter 1 and in section 1 of Chapter 2. To understand this portion of the text no mathematics is needed beyond what is usually covered in a year and a half of high school algebra. For the rest of the booklet, however, it would be desirable to have the background of two years of algebra and a course in trigonometry.

This booklet is particularly suitable for independent study. It contains a large number of problems, some of which are solved as part of the text. More than half of the problems, however, are left as exercises for the reader, with solutions given at the end of the booklet.

CONTENTS

Introduction

Any statement can be classified as *general* or *particular*.
Examples of *general* statements are:

All citizens of the U.S.S.R. are entitled to an education.

The diagonals of a parallelogram bisect one another.

All numbers ending in zero are divisible by 5.

Examples of *particular* statements are:

Petrov is entitled to an education.

The diagonals of the parallelogram $ABCD$ bisect one another.

140 is divisible by 5.

The process of deriving a particular statement from a general statement is called *deduction.* Consider these statements:

(1) All citizens of the U.S.S.R. are entitled to an education.

(2) Petrov is a citizen of the U.S.S.R.

(3) Petrov is entitled to an education.

From general statement (1), together with particular statement (2), we obtain particular statement (3).

The process of obtaining general statements from particular statements is called *induction.* Inductive reasoning may lead to false as well as to true conclusions. We shall clarify this point by two examples.

FIRST EXAMPLE.

(1) 140 is divisible by 5.

(2) All numbers ending in zero are divisible by 5.

General statement (2), obtained from particular statement (1), is true.

SECOND EXAMPLE.

(1) 140 is divisible by 5.

(2) All three-place numbers are divisible by 5.

In this case, general statement (2), derived from particular statement (1), is false.

How can induction be used in mathematics so that only true general statements are obtained from particular statements? The answer to this question is given in this booklet.

1. The Method of Mathematical Induction

1. EXAMPLES OF UNSOUND INDUCTION IN MATHEMATICS

First let us consider two examples of the sort of induction which is inadmissible in mathematics.

EXAMPLE 1.   Let

$$S_n = \frac{1}{1\cdot 2} + \frac{1}{2\cdot 3} + \frac{1}{3\cdot 4} + \cdots + \frac{1}{n(n+1)}.$$

The following equalities are easy to verify:

$$S_1 = \frac{1}{1\cdot 2} = \frac{1}{2},$$

$$S_2 = \frac{1}{1\cdot 2} + \frac{1}{2\cdot 3} = \frac{2}{3},$$

$$S_3 = \frac{1}{1\cdot 2} + \frac{1}{2\cdot 3} + \frac{1}{3\cdot 4} = \frac{3}{4},$$

$$S_4 = \frac{1}{1\cdot 2} + \frac{1}{2\cdot 3} + \frac{1}{3\cdot 4} + \frac{1}{4\cdot 5} = \frac{4}{5}.$$

On the basis of these four results, one might conclude that for *all* natural numbers n it is true that

$$S_n = \frac{n}{n+1}.$$

EXAMPLE 2.   Next let us consider the trinomial $x^2 + x + 41$, first considered by the famous mathematician L. Euler.[1] If we replace x by the number 0 in this trinomial, we obtain the *prime* number 41. If we replace x by the number 1, we again obtain a *prime* number, namely 43. If we continue in this manner and replace x by the numbers 2, 3, 4, 5, 6, 7, 8, 9, 10, successively, we obtain *in each case a prime number*: 47, 53, 61, 71, 83, 97, 113, 131, 151, respectively. On the basis of these results one might conclude that for *any* nonnegative integer x, the value of the trinomial is a *prime number*.

[1] Swiss mathematician and physicist (1707–1783).

3

Why is the reasoning used in these examples inadmissible in mathematics? In what way is the reasoning invalid?

In both the examples the reasoning employed led us to assert a *general statement* referring to *any n* (or any *x*) on the basis that the statements had been found true for certain *particular* values of *n* (or of *x*). It so happens that the general statement obtained in Example 1 is true, as we shall see below in Example 5; however, the general statement obtained in Example 2 is false. Indeed, while it can be shown that the trinomial $x^2 + x + 41$ yields prime numbers for $x = 0, 1, 2, 3, \ldots, 39$, for $x = 40$ the value of the trinomial is seen to be 41^2, which is a composite number.

Induction has wide applications in mathematics, but it must be used with care or it may lead to erroneous conclusions.

2. MORE EXAMPLES OF UNSOUND INDUCTION

We have encountered in Example 2 a statement which proves to be true in forty instances, but which is not true in general. We shall give two additional examples of statements which are true in some particular instances without being true in general.

EXAMPLE 3. The binomial $x^n - 1$ (where *n* is a natural number) is of great interest to mathematicians as it is closely related to the geometric problem of dividing a circle into *n* equal parts. Hence, it is not surprising that this binomial has been studied with great care, with particular attention to the question of resolving it into factors with integral coefficients.

In studying these factorizations for many particular values of *n*, mathematicians noted that in each of the cases studied, the absolute values of the coefficients of the factors never exceeded the number 1. Thus,

$$x - 1 = x - 1,$$
$$x^2 - 1 = (x - 1)(x + 1),$$
$$x^3 - 1 = (x - 1)(x^2 + x + 1),$$
$$x^4 - 1 = (x - 1)(x + 1)(x^2 + 1),$$
$$x^5 - 1 = (x - 1)(x^4 + x^3 + x^2 + x + 1),$$
$$x^6 - 1 = (x - 1)(x + 1)(x^2 + x + 1)(x^2 - x + 1),$$

. . . .

Tables of coefficients were made, and in each case the coefficients had the property mentioned above. Nevertheless, all attempts to prove the statement true for all values of n failed.

The problem was finally solved in 1941 by V. Ivanov with the following result: If $n < 105$, the binomial $x^n - 1$ has the above property. One of the factors of $x^{105} - 1$, however, is the polynomial

$$x^{48} + x^{47} + x^{46} - x^{43} - x^{42} - 2x^{41} - x^{40} - x^{39} + x^{36} + x^{35} + x^{34}$$
$$+ x^{33} + x^{32} + x^{31} - x^{28} - x^{26} - x^{24} - x^{22} - x^{20} + x^{17} + x^{16}$$
$$+ x^{15} + x^{14} + x^{13} + x^{12} - x^9 - x^8 - 2x^7 - x^6 - x^5 + x^2$$
$$+ x + 1,$$

which no longer has this property.

EXAMPLE 4. Into how many parts is space divided by n planes which pass through the same point, but no three of which intersect in the same straight line?

Let us consider the simplest special cases of this problem: One plane divides space into 2 parts. Two planes passing through a common point divide space into 4 parts. Three planes passing through a common point, but having no line in common, divide space into 8 parts.

At first glance it may appear that as the number of planes increases by 1, the number of parts into which they divide space is doubled, so that 4 planes would divide space into 16 parts, 5 planes into 32 parts, and so on; or, in general, that n planes would divide space into 2^n parts.

Actually, this is not the case: 4 planes divide space into 14 parts, 5 planes into 22 parts. In general, n planes divide space into $n(n - 1) + 2$ parts.[1]

These examples illustrate the following simple but important fact: *A statement may be true in many special instances and yet not be true in general.*

3. THE PRINCIPLE OF MATHEMATICAL INDUCTION

Now the following question arises: Suppose that we have a statement involving natural numbers which we have found to be true in some particular instances. How can we determine whether

[1] Proved below in Problem 30, page 21.

the statement is true in general; or, to be more specific, how can we determine whether the statement is true for all natural numbers $n = 1, 2, 3, \ldots$?

This question can sometimes be answered by using a special method of reasoning called the *method of mathematical induction* (complete induction), based on the following principle:

A statement is true for every natural number n if the following conditions are satisfied:

Condition 1.   The statement is true for $n = 1$.

Condition 2.   The truth of the statement for any natural number $n = k$ implies its truth for the succeeding natural number $n = k + 1$.

Proof. We want to show that if conditions 1 and 2 hold, then the statement must be true for every natural number n. We give an indirect proof: If it were not true for every natural number, there would be a smallest natural number, call it m, for which the statement is false. By condition 1, $m \neq 1$. Thus, $m > 1$, so that $m - 1$ is a natural number. Recalling that m is the smallest natural number for which the statement is false, we see that the statement is true for $m - 1$ but false for the succeeding natural number $(m - 1) + 1 = m$. But this contradicts condition 2. Therefore, the statement must be true for every natural number.

Remark. In our proof of the principle of mathematical induction, we used the property of natural numbers that every set of natural numbers contains a smallest number. One can easily show that this property in turn can be deduced from the principle of mathematical induction. Hence, the two propositions are equivalent: Either one can be taken as an *axiom* defining the sequence of natural numbers; the other statement is then a theorem.

4. PROOF BY MATHEMATICAL INDUCTION

A proof by mathematical induction necessarily consists of two parts, that is, proof that the statement satisfies the two independent conditions given in section 3.

EXAMPLE 5.   Find the sum (see Example 1)

$$S_n = \frac{1}{1 \cdot 2} + \frac{1}{2 \cdot 3} + \frac{1}{3 \cdot 4} + \cdots + \frac{1}{n(n + 1)}.$$

SOLUTION.   We know that

$$S_1 = \frac{1}{2}, \quad S_2 = \frac{2}{3}, \quad S_3 = \frac{3}{4}, \quad S_4 = \frac{4}{5}.$$

This time we shall not repeat the erroneous reasoning used in Example 1, and we shall not jump to the conclusion that for all natural numbers n

$$S_n = \frac{n}{n+1}.$$

Rather, we shall be cautious and say only that the results for S_1, S_2, S_3, and S_4 suggest the *hypothesis* that

$$S_n = \frac{n}{n+1}$$

holds for all natural numbers n. To test this hypothesis, we shall use the method of mathematical induction.

Condition 1. The hypothesis is true for $n = 1$, as shown above.

Condition 2. Let us assume that the hypothesis is true for $n = k$, that is, that

$$S_k = \frac{1}{1\cdot 2} + \frac{1}{2\cdot 3} + \cdots + \frac{1}{k(k+1)} = \frac{k}{k+1},$$

where k is any natural number. Let us prove that, as a consequence, the hypothesis must also hold for $n = k + 1$, that is, that

$$S_{k+1} = \frac{k+1}{k+2}.$$

But

$$S_{k+1} = S_k + \frac{1}{(k+1)(k+2)},$$

and on substituting the assumed value for S_k, we have

$$S_{k+1} = \frac{k}{k+1} + \frac{1}{(k+1)(k+2)} = \frac{k^2 + 2k + 1}{(k+1)(k+2)} = \frac{k+1}{k+2}.$$

Thus, both conditions are satisfied. Hence, on the basis of the principle of mathematical induction, we may now assert that for *every natural number n*,

$$S_n = \frac{n}{n+1}.$$

First Remark. We must emphasize that a proof by mathematical induction necessarily requires proofs of both condition 1 and condition 2. We have already seen (Example 2) what erroneous results can arise when condition 2 is disregarded. The following example shows that condition 1 may not be disregarded either.

EXAMPLE 6. "Theorem": *Every natural number is equal to the number which follows it.*

"Proof": Assuming that

$$k = k + 1, \tag{1}$$

let us prove that

$$k + 1 = k + 2. \tag{2}$$

Indeed, by adding 1 to both sides of equality (1), we get equality (2). Hence, if the statement in the "theorem" is true for $n = k$, it is also true for $n = k + 1$; the "theorem" is thus "proved."

"Corollary": All natural numbers are equal to each other.

Wherein did we *err*? We erred in proving only condition 2 and not condition 1. Proof of condition 1 is indispensable in the application of the principle of mathematical induction, and in the present case, condition 1 is not satisfied.

Each of conditions 1 and 2 has its own special significance. Condition 1, so to speak, provides the basis for the induction; condition 2 furnishes the justification for generalizing from this basis— the justification for passing from a special instance to the next, from n to $n + 1$. If condition 1 is not satisfied, then there is no basis for applying the method of mathematical induction—even if condition 2 is satisfied (see Example 6). Where, on the other hand, only the first condition is satisfied and not the second (see solutions of Examples 1 and 2), even though a basis is provided for the induction, there is no justification for making the generalization.

Second Remark. The examples above represent very simple cases of the uses of the method of mathematical induction. In more complicated cases the formulation of the two conditions has to be altered accordingly:

(a) Sometimes to prove that the statement is true for $n = k + 1$, we need to know that the statement is true for $n = k$ and $n = k - 1$, that is, the two numbers preceding $k + 1$. In such cases, in place of condition 1 we would need to prove the assertion for two consecutive values of n (see Chapter 2, Problem 18).

(b) Sometimes we wish to prove a statement for all values of n

greater than or equal to some integer m. In such cases one verifies in the first part of the proof that the statement is true for $n = m$, and, if necessary, for certain larger values of n (see Chapter 2, Problem 24). For example, we often encounter statements that are to be proved for all nonnegative values of n (that is, for $n \geq 0$). In such cases we show in the first part of the proof that the statement is true for $n = 0$; in the second part we proceed as usual.

5. USING MATHEMATICAL INDUCTION TO DETECT FAULTY CONJECTURES

In concluding this chapter, we return once more to Example 1 in order to clarify an essential point concerning the method of mathematical induction.

When we considered the sum

$$S_n = \frac{1}{1 \cdot 2} + \frac{1}{2 \cdot 3} + \cdots + \frac{1}{n(n + 1)}$$

for various values of n, we found that

$$S_1 = \frac{1}{2}, \quad S_2 = \frac{2}{3}, \quad S_3 = \frac{3}{4}, \quad S_4 = \frac{4}{5}.$$

This led to the hypothesis that for any value of n

$$S_n = \frac{n}{n + 1}.$$

To prove this hypothesis, we used the method of mathematical induction in Example 5. We were lucky in that we found our hypothesis to be correct. If instead we had hit upon an incorrect hypothesis, its falseness would have been revealed in the second part of the proof, as shown in the following example.

EXAMPLE 7. We know that

$$S_n = \frac{1}{1 \cdot 2} + \frac{1}{2 \cdot 3} + \cdots + \frac{1}{n(n + 1)} = \frac{n}{n + 1}. \tag{1}$$

However, let us suppose that in studying S_n, we had stated the hypothesis

$$S_n = \frac{n + 1}{3n + 1}. \tag{2}$$

Formula (2) works for $n = 1$, since $S_1 = \frac{1}{2}$. Assuming that formula (2) holds for $n = k$ would mean that

$$S_k = \frac{k+1}{3k+1}.$$

Let us try to prove that, if true for $n = k$, formula (2) would also be true for $n = k + 1$, that is, that

$$S_{k+1} = \frac{k+2}{3k+4}.$$

But

$$S_{k+1} = S_k + \frac{1}{(k+1)(k+2)}$$

$$= \frac{k+1}{3k+1} + \frac{1}{(k+1)(k+2)} = \frac{k^3 + 4k^2 + 8k + 3}{(k+1)(k+2)(3k+1)},$$

which is not the expected result. Thus, the validity of the formula for $n = k$ does not imply its validity for $n = k + 1$. We have discovered that formula (2) is false.

Thus, we see that the method of mathematical induction makes it possible to prove correct generalizations and to reject false ones.

2. Examples and Exercises

To learn to use the method of mathematical induction, one has to solve a good many problems. This chapter contains 52 problems. In 22 of them solutions are given in detail in the text. The solutions of the remaining 30, which are meant to be solved by the reader, are given at the end of the booklet.

6. ELEMENTARY EXAMPLES AND EXERCISES FROM ALGEBRA AND GEOMETRY

PROBLEM 1. Let us write the sequence of odd numbers in order of magnitude: 1, 3, 5, 7, ... Let us designate the first by u_1, the second by u_2, the third by u_3, etc.; that is, we set

$$u_1 = 1, \quad u_2 = 3, \quad u_3 = 5, \quad u_4 = 7, \quad \ldots.$$

Let us now pose the problem of finding a formula which expresses the odd number u_n in terms of its index n.

SOLUTION. The first odd number u_1 can be written in the form

$$u_1 = 2 \cdot 1 - 1; \tag{1}$$

the second odd number u_2 can be written in the form

$$u_2 = 2 \cdot 2 - 1; \tag{2}$$

the third odd number u_3 can be written in the form

$$u_3 = 2 \cdot 3 - 1. \tag{3}$$

Careful examination of equalities (1), (2), and (3) leads to the hypothesis that any odd number can be obtained by multiplying its index by 2 and subtracting 1; that is, that for any odd number u_n,

$$u_n = 2n - 1. \tag{4}$$

Let us prove that formula (4) is generally valid.

Condition 1. Equality (1) shows that formula (4) holds for $n = 1$.

Condition 2. Let us assume that formula (4) holds for $n = k$; that is, that the kth odd number is given by

$$u_k = 2k - 1.$$

Let us show that, on this assumption, formula (4) holds also for the $(k + 1)$st odd number, that is, that the $(k + 1)$st odd number must be given by

$$u_{k+1} = 2(k + 1) - 1, \text{ or, equivalently, } u_{k+1} = 2k + 1.$$

To obtain the $(k + 1)$st odd number, one has only to add 2 to the kth odd number; thus, $u_{k+1} = u_k + 2$. By assumption, $u_k = 2k - 1$. Consequently,

$$u_{k+1} = (2k - 1) + 2 = 2k + 1.$$

Result. $u_n = 2n - 1$.

PROBLEM 2. Compute the sum of the first n odd numbers.

SOLUTION. Let us denote the required sum by S_n. Thus

$$S_n = 1 + 3 + 5 + \cdots + (2n - 1).$$

This kind of problem can be solved by using a ready-made formula. But we are interested in solving the problem without resorting to such a formula, by using the method of mathematical induction. To do this, it is necessary first to state a hypothesis, that is, simply to try to guess the solution.

We assign to n the successive values 1, 2, 3, ... until there is enough information to suggest a promising hypothesis. Then we have to test this hypothesis by the method of mathematical induction. We find

$$S_1 = 1, \quad S_2 = 4, \quad S_3 = 9, \quad S_4 = 16, \quad S_5 = 25, \quad S_6 = 36.$$

Now everything depends on the powers of observation of the student—on his ability to conjecture a general relation from the particular results. In the cases given above, it is at once apparent that

$$S_1 = 1^2, \quad S_2 = 2^2, \quad S_3 = 3^2, \quad S_4 = 4^2.$$

From this it is reasonable to suppose that, in general,

$$S_n = n^2.$$

Let us prove this hypothesis.

Condition 1. For $n = 1$ the sum consists of a single term, namely, the number 1. For $n = 1$ the value of the expression n^2 is also 1. Hence, our hypothesis holds for $n = 1$.

Condition 2. Let us assume that the hypothesis holds for $n = k$, that is, that $S_k = k^2$. Let us prove that on this assumption the hypothesis must also hold for $n = k + 1$, that is, that

$$S_{k+1} = (k + 1)^2.$$

Indeed, since $u_{k+1} = 2k + 1$ from Problem 1,

$$S_{k+1} = S_k + (2k + 1).$$

But $S_k = k^2$, and hence,

$$S_{k+1} = k^2 + (2k + 1) = (k + 1)^2.$$

Result. $S_n = n^2$.

PROBLEM 3. Find the general term of the sequence of numbers u_n if $u_1 = 1$ and if for every natural number $k > 1$ the relation $u_k = u_{k-1} + 3$ holds.

Hint. $u_1 = 3 \cdot 1 - 2$; $u_2 = 3 \cdot 2 - 2$.

PROBLEM 4. Find the sum

$$S_n = 1 + 2 + 2^2 + 2^3 + \cdots + 2^{n-1}.$$

Hint. (1) $S_1 = 2 - 1$; $S_2 = 2^2 - 1$; or (2) consider $2S_n - S_n$.

PROBLEM 5. Prove that the sum S_n of the first n natural numbers is

$$\frac{n(n + 1)}{2}.$$

SOLUTION. This problem differs from the preceding ones in that it is not necessary to look for a hypothesis; the hypothesis is given. We merely have to prove that it is correct.

$$S_n = 1 + 2 + 3 + 4 + \cdots + n.$$

Condition 1. The hypothesis holds for $n = 1$.
Condition 2. Assuming that

$$S_k = 1 + 2 + \cdots + k = \frac{k(k + 1)}{2},$$

let us show that $$S_{k+1} = \frac{(k + 1)(k + 2)}{2}.$$

Indeed,

$$S_{k+1} = S_k + (k + 1) = \frac{k(k + 1)}{2} + (k + 1) = \frac{(k + 1)(k + 2)}{2}.$$

This completes the proof.

PROBLEM 6. Prove that the sum of the squares of the first n natural numbers is $\dfrac{n(n + 1)(2n + 1)}{6}$.

PROBLEM 7. Prove that

$$S_n = 1^2 - 2^2 + 3^2 - 4^2 + \cdots + (-1)^{n-1}n^2 = (-1)^{n-1}\frac{n(n + 1)}{2}.$$

SOLUTION. *Condition 1.* The hypothesis obviously holds for $n = 1$, since $(-1)^0 = 1$.

Condition 2. Assuming that

$$S_k = 1 - 2^2 + 3^2 - \cdots + (-1)^{k-1}k^2 = (-1)^{k-1}\frac{k(k + 1)}{2},$$

let us prove that

$$S_{k+1} = 1 - 2^2 + 3^2 - \cdots + (-1)^{k-1}k^2 + (-1)^k(k + 1)^2$$
$$= (-1)^k\frac{(k + 1)(k + 2)}{2}.$$

Indeed,

$$S_{k+1} = S_k + (-1)^k(k + 1)^2$$
$$= (-1)^{k-1}\frac{k(k + 1)}{2} + (-1)^k(k + 1)^2$$
$$= (-1)^k\left[(k + 1) - \frac{k}{2}\right](k + 1) = (-1)^k\frac{(k + 1)(k + 2)}{2}.$$

PROBLEM 8. Prove that

$$1^2 + 3^2 + 5^2 + \cdots + (2n - 1)^2 = \frac{n(2n - 1)(2n + 1)}{3}.$$

PROBLEM 9. Prove that the sum of the cubes of the first n natural numbers is $\left[\dfrac{n(n + 1)}{2}\right]^2$.

PROBLEM 10. Prove that

$$1 + x + x^2 + \cdots + x^n = \frac{x^{n+1} - 1}{x - 1} \qquad (x \neq 1).$$

PROBLEM 11. Prove that

$$1 \cdot 2 + 2 \cdot 3 + 3 \cdot 4 + \cdots + n(n + 1) = \frac{n(n + 1)(n + 2)}{3}.$$

PROBLEM 12. Prove that

$$1 \cdot 2 \cdot 3 + 2 \cdot 3 \cdot 4 + 3 \cdot 4 \cdot 5 + \cdots + n(n + 1)(n + 2)$$
$$= \frac{n(n + 1)(n + 2)(n + 3)}{4}.$$

PROBLEM 13. Prove that

$$\frac{1}{1 \cdot 3} + \frac{1}{3 \cdot 5} + \cdots + \frac{1}{(2n - 1)(2n + 1)} = \frac{n}{2n + 1}.$$

PROBLEM 14. Prove that

$$\frac{1^2}{1 \cdot 3} + \frac{2^2}{3 \cdot 5} + \cdots + \frac{n^2}{(2n - 1)(2n + 1)} = \frac{n(n + 1)}{2(2n + 1)}.$$

PROBLEM 15. Prove that

$$\frac{1}{1 \cdot 4} + \frac{1}{4 \cdot 7} + \frac{1}{7 \cdot 10} + \cdots + \frac{1}{(3n - 2)(3n + 1)} = \frac{n}{3n + 1}.$$

PROBLEM 16. Prove that

$$\frac{1}{1 \cdot 5} + \frac{1}{5 \cdot 9} + \frac{1}{9 \cdot 13} + \cdots + \frac{1}{(4n - 3)(4n + 1)} = \frac{n}{4n + 1}.$$

PROBLEM 17. Prove that

$$\frac{1}{a(a + 1)} + \frac{1}{(a + 1)(a + 2)} + \cdots + \frac{1}{(a + n - 1)(a + n)}$$
$$= \frac{n}{a(a + n)}.$$

PROBLEM 18. Prove that for all nonnegative integers n (that is, for $n \geq 0$)

$$v_n = 2^n + 1,$$

given that (i) $v_0 = 2$, $v_1 = 3$, and (ii) for every natural number k, the relation $v_{k+1} = 3v_k - 2v_{k-1}$ holds.

SOLUTION. It is clear from what is **given** that the statement holds for $n = 0$ and for $n = 1$. (In this connection see the Second Remark of section 4 in Chapter 1.) Let us assume that

$$v_{k-1} = 2^{k-1} + 1; \quad v_k = 2^k + 1.$$

It then follows that

$$v_{k+1} = 3(2^k + 1) - 2(2^{k-1} + 1) = 2^{k+1} + 1.$$

PROBLEM 19. Prove that

$$u_n = \frac{\alpha^{n+1} - \beta^{n+1}}{\alpha - \beta},$$

if

$$u_1 = \frac{\alpha^2 - \beta^2}{\alpha - \beta}, \quad u_2 = \frac{\alpha^3 - \beta^3}{\alpha - \beta} \qquad (\alpha \neq \beta),$$

and if for every natural number $k > 2$ the following relation holds:

$$u_k = (\alpha + \beta)u_{k-1} - \alpha\beta u_{k-2}.$$

PROBLEM 20. The product $1 \cdot 2 \cdot 3 \cdot \cdots \cdot n$ is designated by the symbol $n!$ (read "n factorial"). Note that $1! = 1$, $2! = 2$, $3! = 6$. $4! = 24$, $5! = 120$. Find an expression for the sum

$$S_n = 1 \cdot 1! + 2 \cdot 2! + 3 \cdot 3! + \cdots + n \cdot n!.$$

SOLUTION. $S_1 = 1 \cdot 1! = 1,$
$$S_2 = 1 \cdot 1! + 2 \cdot 2! = 5,$$
$$S_3 = 1 \cdot 1! + 2 \cdot 2! + 3 \cdot 3! = 23,$$
$$S_4 = 1 \cdot 1! + 2 \cdot 2! + 3 \cdot 3! + 4 \cdot 4! = 119.$$

Examining these sums, one notes that

$$S_1 = 2! - 1, \quad S_2 = 3! - 1, \quad S_3 = 4! - 1, \quad S_4 = 5! - 1.$$

This suggests the hypothesis

$$S_n = (n + 1)! - 1.$$

Let us verify this hypothesis.

Condition 1. The hypothesis holds for $n = 1$, since

$$S_1 = 1 \cdot 1! = 2! - 1.$$

Condition 2. Assuming that
$$S_k = 1 \cdot 1! + 2 \cdot 2! + \cdots + k \cdot k! = (k + 1)! - 1,$$
let us show that
$$S_{k+1} = 1 \cdot 1! + 2 \cdot 2! + \cdots + k \cdot k! + (k + 1) \cdot (k + 1)!$$
$$= (k + 2)! - 1.$$

Indeed,
$$\begin{aligned} S_{k+1} &= S_k + (k + 1) \cdot (k + 1)! \\ &= [(k + 1)! - 1] + (k + 1) \cdot (k + 1)! \\ &= (k + 1)! [1 + (k + 1)] - 1 \\ &= (k + 1)!(k + 2) - 1 = (k + 2)! - 1. \end{aligned}$$

PROBLEM 21. Prove that for all nonnegative integers n, the following identity holds:
$$\frac{1}{1 + x} + \frac{2}{1 + x^2} + \frac{4}{1 + x^4} + \frac{8}{1 + x^8} + \cdots + \frac{2^n}{1 + x^{2^n}}$$
$$= \frac{1}{x - 1} + \frac{2^{n+1}}{1 - x^{2^{n+1}}} \qquad (x \neq 1, \, x \neq -1).$$

(In connection with this problem, see the Second Remark of section 4 in Chapter 1.)

PROBLEM 22. Given numbers α and β, with $\alpha \neq \beta$ and $\alpha + \beta \neq 1$; let
$$\alpha + \beta = m, \quad \alpha\beta = a, \quad A_2 = m - \frac{a}{m - 1},$$
$$A_3 = m - \frac{a}{m - \dfrac{a}{m - 1}}, \quad A_4 = m - \frac{a}{m - \dfrac{a}{m - \dfrac{a}{m - 1}}}, \quad \text{etc.,}$$
that is, for every $k > 1$,
$$A_{k+1} = m - \frac{a}{A_k}.$$
Prove that
$$A_n = \frac{(\alpha^{n+1} - \beta^{n+1}) - (\alpha^n - \beta^n)}{(\alpha^n - \beta^n) - (\alpha^{n-1} - \beta^{n-1})}. \qquad (1)$$

SOLUTION. *Condition 1.* First, let us show that formula (1) holds for $n = 2$. By assumption,

$$A_2 = m - \frac{a}{m - 1} = (\alpha + \beta) - \frac{\alpha\beta}{(\alpha + \beta) - 1}$$

$$= \frac{\alpha^2 + \beta^2 + \alpha\beta - \alpha - \beta}{\alpha + \beta - 1}.$$

According to formula (1),

$$A_2 = \frac{(\alpha^3 - \beta^3) - (\alpha^2 - \beta^2)}{(\alpha^2 - \beta^2) - (\alpha - \beta)}.$$

Noting that $\alpha - \beta$ is a factor of both the numerator and the denominator of the above fraction, we may reduce the fraction to obtain

$$A_2 = \frac{\alpha^2 + \beta^2 + \alpha\beta - \alpha - \beta}{\alpha + \beta - 1}.$$

Condition 2. Assuming that formula (1) holds for $n = k$, let us prove that it must also be true for $n = k + 1$. That is, if

$$A_k = \frac{(\alpha^{k+1} - \beta^{k+1}) - (\alpha^k - \beta^k)}{(\alpha^k - \beta^k) - (\alpha^{k-1} - \beta^{k-1})}, \tag{2}$$

then

$$A_{k+1} = \frac{(\alpha^{k+2} - \beta^{k+2}) - (\alpha^{k+1} - \beta^{k+1})}{(\alpha^{k+1} - \beta^{k+1}) - (\alpha^k - \beta^k)}.$$

Indeed,

$$A_{k+1} = m - \frac{a}{A_k} \quad \text{or} \quad A_{k+1} = (\alpha + \beta) - \frac{\alpha\beta}{A_k}.$$

Making use of equality (2) we get

$$A_{k+1} = (\alpha + \beta) - \frac{\alpha\beta[(\alpha^k - \beta^k) - (\alpha^{k-1} - \beta^{k-1})]}{(\alpha^{k+1} - \beta^{k+1}) - (\alpha^k - \beta^k)}$$

$$= \frac{(\alpha^{k+2} - \beta^{k+2}) - (\alpha^{k+1} - \beta^{k+1})}{(\alpha^{k+1} - \beta^{k+1}) - (\alpha^k - \beta^k)}.$$

This completes the proof.

PROBLEM 23. Simplify the polynomial

$$1 - \frac{x}{1!} + \frac{x(x-1)}{2!} - \cdots + (-1)^n \frac{x(x-1)\cdots(x-n+1)}{n!}.$$

Answer. $(-1)^n \dfrac{(x-1)(x-2)\cdots(x-n)}{n!}$. The reader should supply the proof.

PROBLEM 24. Prove that any whole number of rubles greater than 7 can be paid with 3-ruble and 5-ruble bills without requiring change.

SOLUTION. The statement holds for 8 rubles. Suppose the statement holds for some natural number $k \geq 8$ rubles.

There are two possibilities: (1) the k rubles may be paid in 3-ruble bills only, or (2) there may be at least one 5-ruble bill among the bills.

In the first case, the number of 3-ruble bills can be no less than three, for $k \geq 8$. Hence, to pay $k + 1$ rubles, three 3-ruble bills can be replaced by two 5-ruble bills.

In the second case, one 5-ruble bill can be replaced by two 3-ruble bills to pay $k + 1$ rubles.

PROBLEM 25. Prove that the sum of the cubes of any three consecutive natural numbers is divisible by 9.

SOLUTION. The sum $1^3 + 2^3 + 3^3$ is divisible by 9. Hence, the statement holds if the first of the three consecutive natural numbers is 1.

Suppose the sum $k^3 + (k+1)^3 + (k+2)^3$ is divisible by 9, where k is some natural number. Then the sum

$$(k+1)^3 + (k+2)^3 + (k+3)^3$$
$$= [k^3 + (k+1)^3 + (k+2)^3] + 9(k^2 + 3k + 3)$$

is the sum of two expressions, each of which is divisible by 9; hence their sum is also divisible by 9.

PROBLEM 26. Prove that for every nonnegative whole number n the sum

$$A_n = 11^{n+2} + 12^{2n+1}$$

is divisible by 133.

PROBLEM 27. From among the $2n$ numbers $1, 2, \ldots, 2n$ choose at random $n + 1$ numbers. Prove that among the numbers chosen there are at least two numbers such that one of them is divisible by the other.

SOLUTION.[1] Let us assume that from among the $2n$ numbers $1, 2, \ldots, 2n$, where $n \geq 2$, we have found $n + 1$ numbers such that no one of them is divisible by any other. Let us denote this set of $n + 1$ numbers by M_{n+1}. Let us prove that it would then be possible to select from among the $2n - 2$ numbers $1, 2, \ldots, 2n - 2$ a set containing n numbers such that no one of the n numbers is divisible by any other.

There are four possibilities:
1. M_{n+1} contains neither the number $2n - 1$ nor the number $2n$.
2. M_{n+1} contains $2n - 1$ but not $2n$.
3. M_{n+1} contains $2n$ but not $2n - 1$.
4. M_{n+1} contains both $2n - 1$ and $2n$.

Case 1. Let us remove an arbitrary number from the set M_{n+1}. There then remain n numbers none of which is greater than $2n - 2$. No one of these is divisible by any other.

Case 2. Let us remove the number $2n - 1$ from the set M_{n+1}. Again, among the remaining n numbers, none is greater than $2n - 2$, and no one of them is divisible by any other.

Case 3. Let us remove the number $2n$ from the set M_{n+1}; the result is the same as in cases 1 and 2.

Case 4. We note first of all that the number n cannot belong to the set M_{n+1}; otherwise the set M_{n+1} would contain the two numbers n and $2n$, and $2n$ is divisible by n.

Now let us remove the two numbers $2n - 1$ and $2n$ from the set M_{n+1}. Let us denote the set of the $n - 1$ numbers which remain by M_{n-1}. Next we adjoin the number n to the set M_{n-1}, thereby obtaining a set of n numbers, no one of which is greater than $2n - 2$. It remains to be shown that of these n numbers none would be divisible by any other.

Since the set M_{n+1} contained no two numbers of which one was divisible by another, the set M_{n-1} contains no such numbers, either. Hence, we have only to make certain that there cannot be any two such numbers even when the number n is added to the set M_{n-1}.

[1] This solution was proposed by a student at the Herzen Leningrad Pedagogical Institute.

To do this, it is sufficient to show (1) that no number in M_{n-1} is divisible by n, and (2) that n is divisible by no number in M_{n-1}.

The first statement follows from the fact that of the numbers in M_{n-1} none is greater than $2n - 2$. The second follows from the fact that $2n$ is not divisible by any number in M_{n-1}. Thus, we have shown that if the statement is false for the $2n$ numbers $1, 2, \ldots, 2n$, it must be false also for the $2(n - 1)$ numbers $1, 2, \ldots, 2n - 2$. Hence, if the statement is true for the $2(n - 1)$ numbers $1, 2, \ldots, 2n - 2$, it must be true also for the $2n$ numbers $1, 2, \ldots, 2n$.

The statement is true for the two numbers 1 and 2; hence, it is true for all sets of $2n$ numbers $1, 2, \ldots, 2n$, where n is a natural number.

Remark. The problem has the following simple solution which does not use the method of mathematical induction. Let us choose any $n + 1$ numbers from the set of $2n$ numbers $1, 2, \ldots, 2n$, and designate this set by M_{n+1}. Let us divide each even number in M_{n+1} by a power of 2 so that the quotient will be an odd number. From these quotients and the odd numbers in M_{n+1} let us form the set $\overline{M}_{n+1}$. $\overline{M}_{n+1}$ contains $n + 1$ odd numbers, each less than $2n$. Since there are only n positive odd numbers less than $2n$, at least two of the quotients in the set $\overline{M}_{n+1}$ must be the same number; let us call this number k. Thus, in the set M_{n+1} there must be at least two numbers of the form $2^s k$ and $2^t k$. It is obvious that one of these two numbers will be divisible by the other.

PROBLEM 28. Prove that n different straight lines, lying in a plane and passing through a common point, divide the plane into $2n$ parts.

PROBLEM 29. Prove that n straight lines which lie in a plane divide the plane into regions which can be colored black and white in such a way that any two neighboring regions (that is, regions which border on a common line segment) have different colors.

7. ADVANCED EXAMPLES AND EXERCISES

PROBLEM 30. Prove that n planes which pass through a common point, no three of which intersect in the same straight line, divide space into $A_n = n(n - 1) + 2$ parts.

SOLUTION. 1) One plane divides space into two parts, and $A_1 = 2$. Hence, the statement holds for $n = 1$.

2) Let us assume that the statement holds for $n = k$, that is, that k planes divide space into $k(k - 1) + 2$ parts. Let us prove that $k + 1$ planes must then divide space into $k(k + 1) + 2$ parts.

Let P be the $(k + 1)$st plane. P intersects each of the other planes in a straight line, and these k straight lines all intersect in the point which is common to all the planes. Using Problem 28, we see that the plane P is divided into $2k$ parts, each of which is enclosed by a plane angle with its vertex at the common point of intersection. The first k planes divide space into several polyhedral angles. Plane P divides some of these polyhedral angles into 2 parts. The face common to two such parts is that portion of the plane bounded by the two half-lines along which P intersects the faces of the given polyhedral angle, that is, one of the $2k$ plane angles into which the plane P is divided. This means that the number of polyhedral angles which are cut into two parts by plane P cannot exceed $2k$.

On the other hand, each of the two parts into which plane P is divided as a result of its intersection with the first plane thus divides the polyhedral angle formed by the first k planes into two parts. This means that the number of polyhedral angles which are cut in two by plane P cannot be less than $2k$.

Thus, plane P divides exactly $2k$ of the polyhedral angles formed by the first k planes into 2 parts. Consequently, if k planes divide space into $k(k - 1) + 2$ parts, then $k + 1$ planes divide space into

$$[k(k - 1) + 2] + 2k = k(k + 1) + 2$$

parts. The statement has thus been proved.

PROBLEM 31.   Prove the identity

$$\cos \alpha \cos 2\alpha \cos 4\alpha \cdots \cos 2^n\alpha = \frac{\sin 2^{n+1}\alpha}{2^{n+1} \sin \alpha}.$$

SOLUTION.   1)   The formula holds for $n = 0$, since

$$\cos \alpha = \frac{\sin 2\alpha}{2 \sin \alpha}.$$

2)   Let us assume that the formula holds for $n = k$, that is, that

$$\cos \alpha \cos 2\alpha \cdots \cos 2^k\alpha = \frac{\sin 2^{k+1}\alpha}{2^{k+1} \sin \alpha}.$$

Then we must show that the formula holds also for $n = k + 1$. Indeed,

$$\cos \alpha \cos 2\alpha \cdots \cos 2^k\alpha \cos 2^{k+1}\alpha$$

$$= \frac{\sin 2^{k+1}\alpha \cos 2^{k+1}\alpha}{2^{k+1} \sin \alpha} = \frac{\sin 2^{k+2}\alpha}{2^{k+2} \sin \alpha}.$$

PROBLEM 32. Given that $A_1 = \cos \theta$, $A_2 = \cos 2\theta$, and that for every natural number $k > 2$

$$A_k = 2 A_{k-1} \cos \theta - A_{k-2};$$

prove that

$$A_n = \cos n\theta.$$

SOLUTION. 1) The formula holds for $n = 1$ and $n = 2$.
2) Let us assume that

$$A_{k-2} = \cos (k - 2)\theta, \quad A_{k-1} = \cos (k - 1)\theta.$$

Then

$$A_k = 2 \cos \theta \cos (k - 1)\theta - \cos (k - 2)\theta = \cos k\theta.[1]$$

PROBLEM 33. Prove that

$$\sin x + \sin 2x + \cdots + \sin nx = \frac{\sin \dfrac{n + 1}{2} x}{\sin \dfrac{x}{2}} \sin \frac{nx}{2}.$$

SOLUTION. 1) The formula holds for $n = 1$.
2) Assume that

$$\sin x + \sin 2x + \cdots + \sin kx = \frac{\sin \dfrac{k + 1}{2} x}{\sin \dfrac{x}{2}} \sin \frac{kx}{2}.$$

---

[1] Here we make use of the identity $2 \cos \alpha \cos \beta = \cos (\alpha + \beta) + \cos (\alpha - \beta)$, letting $\alpha = (k - 1)\theta$ and $\beta = \theta$.

Then $\quad \sin x + \sin 2x + \cdots + \sin kx + \sin (k + 1)x$

$$= \frac{\sin \dfrac{k + 1}{2}x}{\sin \dfrac{x}{2}} \sin \frac{kx}{2} + \sin (k + 1)x$$

$$= \frac{\sin \dfrac{k + 1}{2}x}{\sin \dfrac{x}{2}} \sin \frac{kx}{2} + 2 \sin \frac{k + 1}{2}x \cos \frac{k + 1}{2}x$$

$$= \frac{\sin \dfrac{k + 2}{2}x}{\sin \dfrac{x}{2}} \sin \frac{k + 1}{2}x,$$

since $\quad 2 \cos \dfrac{k + 1}{2}x \sin \dfrac{x}{2} = \sin \dfrac{k + 2}{2}x - \sin \dfrac{kx}{2}.$

PROBLEM 34. Prove that

$$\frac{1}{2} + \cos x + \cos 2x + \cdots + \cos nx = \frac{\sin \dfrac{2n + 1}{2}x}{2 \sin \dfrac{x}{2}}.$$

PROBLEM 35. Prove that

$$\sin x + 2 \sin 2x + 3 \sin 3x + \cdots + n \sin nx$$
$$= \frac{(n + 1) \sin nx - n \sin (n + 1)x}{4 \sin^2 \dfrac{x}{2}}.$$

PROBLEM 36. Prove that

$$\cos x + 2 \cos 2x + \cdots + n \cos nx$$
$$= \frac{(n + 1) \cos nx - n \cos (n + 1)x - 1}{4 \sin^2 \dfrac{x}{2}}.$$

PROBLEM 37. Prove that

$$\frac{1}{2} \tan \frac{x}{2} + \frac{1}{2^2} \tan \frac{x}{2^2} + \cdots + \frac{1}{2^n} \tan \frac{x}{2^n}$$
$$= \frac{1}{2^n} \cot \frac{x}{2^n} - \cot x \qquad (x \neq m\pi).$$

PROBLEM 38. Prove that

arc cot 3 + arc cot 5 + $\cdots$ + arc cot $(2n + 1)$

$= \text{arc tan } 2 + \text{arc tan } \dfrac{3}{2} + \cdots + \text{arc tan } \dfrac{n+1}{n} - n \text{ arc tan } 1.$

PROBLEM 39. Prove that

$$(1 + i)^n = 2^{\frac{n}{2}} \left(\cos \frac{n\pi}{4} + i \sin \frac{n\pi}{4} \right).$$

SOLUTION. 1) The equality holds for $n = 1$, since

$$1 + i = 2^{\frac{1}{2}} \left(\cos \frac{\pi}{4} + i \sin \frac{\pi}{4} \right).$$

2) Let us assume that

$$(1 + i)^k = 2^{\frac{k}{2}} \left(\cos \frac{k\pi}{4} + i \sin \frac{k\pi}{4} \right).$$

Then

$$(1 + i)^{k+1} = 2^{\frac{k}{2}} \left(\cos \frac{k\pi}{4} + i \sin \frac{k\pi}{4} \right) \cdot 2^{\frac{1}{2}} \left(\cos \frac{\pi}{4} + i \sin \frac{\pi}{4} \right)$$

$$= 2^{\frac{k+1}{2}} \left(\cos \frac{(k + 1)\pi}{4} + i \sin \frac{(k + 1)\pi}{4} \right).$$

PROBLEM 40. Prove that

$$(\sqrt{3} - i)^n = 2^n \left(\cos \frac{n\pi}{6} - i \sin \frac{n\pi}{6} \right).$$

PROBLEM 41. Prove the following theorem: If the application of a finite number of rational operations (that is, addition, subtraction, multiplication, and division) to the complex numbers $x_1, x_2, \ldots, x_n$ results in the number u, then the application of the same operations in the same order of succession to the complex conjugates $\bar{x}_1, \bar{x}_2, \ldots, \bar{x}_n$ results in the number $\bar{u}$, which is the conjugate of u.

SOLUTION. First let us prove that the theorem holds for each of the four operations applied simply to two complex numbers. Let

$$x_1 = a + bi, \quad x_2 = c + di.$$

Then
$$x_1 + x_2 = (a + c) + (b + d)i = u;$$
$$\bar{x}_1 + \bar{x}_2 = (a - bi) + (c - di) = (a + c) - (b + d)i = \bar{u}.$$

In exactly the same way the theorem can be verified for subtraction, multiplication, and division.

Now let there be given a rational expression involving the complex numbers $x_1, x_2, \ldots, x_n$. As is well known, the simplification of such an expression can be reduced to successive applications of the four rational operations on two complex numbers, and these operations can be applied in a particular order.

For example, let

$$u = \frac{x_1 x_2 + x_3 x_4}{x_1 + x_2 - x_3}.$$

To compute u, it is sufficient to carry out the following operations in the indicated order:

(1) $x_1 x_2 = u_1,$ (4) $u_3 - x_3 = u_4,$

(2) $x_3 x_4 = u_2,$ (5) $u_1 + u_2 = u_5,$

(3) $x_1 + x_2 = u_3,$ (6) $\dfrac{u_5}{u_4} = u.$

Let us assume that the theorem holds for all rational expressions whose simplification requires not more than k applications of the "operations." Let us show that then the theorem must hold also for expressions requiring the application of not more than $k + 1$ applications of the "operations." If the numbers $x_1, x_2, \ldots, x_n$ are replaced by their complex conjugates, the numbers u_i and u_j are replaced by their complex conjugates, $\bar{u}_i$ and $\bar{u}_j$, and the $(k + 1)$st "operation" applied to $\bar{u}_i$ and $\bar{u}_j$ gives the number $\bar{u}$, the complex conjugate of u.

PROBLEM 42. Prove that for every natural number n

$$(\cos x + i \sin x)^n = \cos nx + i \sin nx.$$

PROBLEM 43. Prove that for all natural numbers $n > 1$

$$\frac{1}{n + 1} + \frac{1}{n + 2} + \cdots + \frac{1}{2n} > \frac{13}{24}.$$

SOLUTION. Let us designate the expression on the left side of the inequality by S_n.

1) $S_2 = \dfrac{7}{12} = \dfrac{14}{24}$, so that the inequality holds for $n = 2$.

2) Let us assume that $S_k > \dfrac{13}{24}$ for some number k. Let us prove that then we must also have $S_{k+1} > \dfrac{13}{24}$. We have

$$S_k = \frac{1}{k+1} + \frac{1}{k+2} + \cdots + \frac{1}{2k},$$

$$S_{k+1} = \frac{1}{k+2} + \frac{1}{k+3} + \cdots + \frac{1}{2k} + \frac{1}{2k+1} + \frac{1}{2k+2}.$$

Subtracting S_k from S_{k+1}, we obtain

$$S_{k+1} - S_k = \frac{1}{2k+1} + \frac{1}{2k+2} - \frac{1}{k+1},$$

that is,

$$S_{k+1} - S_k = \frac{1}{2(k+1)(2k+1)}.$$

For any natural number k the expression on the right side of the last equality is positive. Hence, $S_{k+1} > S_k$. But $S_k > \dfrac{13}{24}$; hence, $S_{k+1} > \dfrac{13}{24}$, also.

PROBLEM 44. Find the fallacy in the following "proof."

STATEMENT: For every natural number n the following inequality holds:

$$2^n > 2n + 1.$$

Proof: Let us assume that the inequality holds for some natural number $n = k$; that is, we assume that

$$2^k > 2k + 1. \tag{1}$$

Let us prove that it then holds also for $n = k + 1$; that is,

$$2^{k+1} > 2(k+1) + 1. \tag{2}$$

For any natural number k, 2^k is no smaller than 2. To the left side of inequality (1) let us add 2^k, and to the right side let us add 2. This gives the inequality

$$2^k + 2^k > 2k + 1 + 2,$$

or

$$2^{k+1} > 2(k + 1) + 1.$$

This proves the inequality.

PROBLEM 45. For what natural numbers n does the inequality $2^n > 2n + 1$ hold?

PROBLEM 46. For what natural numbers n does the inequality $2^n > n^2$ hold?
    SOLUTION.

The inequality holds for $n = 1$, since $2^1 > 1^2$.
The inequality does not hold for $n = 2$, since $2^2 = 2^2$.
The inequality does not hold for $n = 3$, since $2^3 < 3^2$.
The inequality does not hold for $n = 4$, since $2^4 = 4^2$.
The inequality holds for $n = 5$, since $2^5 > 5^2$.
The inequality holds for $n = 6$, since $2^6 > 6^2$.

Apparently the inequality holds for $n = 1$, and for all $n > 4$. Let us prove this.
    1) The inequality holds for $n = 5$.
    2) Let us assume that, for some natural number $k > 4$,

$$2^k > k^2. \tag{1}$$

Let us prove that

$$2^{k+1} > (k + 1)^2. \tag{2}$$

We know that $2^k > 2k + 1$ for $k > 4$ (Problem 45). Hence, by adding 2^k to the left side and $2k + 1$ to the right side of inequality (1), we obtain inequality (2).
    Result. $2^n > n^2$ for $n = 1$ and for all $n > 4$.

PROBLEM 47.   Prove that

$$(1 + \alpha)^n > 1 + n\alpha,$$

where $\alpha > -1$, $\alpha \neq 0$, and n is a natural number greater than 1.

SOLUTION.   1)   The inequality holds for $n = 2$, since $\alpha^2 > 0$.

2)   Let us assume that the inequality holds for some natural number $n = k$; that is,

$$(1 + \alpha)^k > 1 + k\alpha. \tag{1}$$

Let us prove that then the inequality holds also for $n = k + 1$; that is,

$$(1 + \alpha)^{k+1} > 1 + (k + 1)\alpha. \tag{2}$$

Since, by our original assumption, $1 + \alpha > 0$, multiplication of both sides of inequality (1) by $1 + \alpha$ gives the valid inequality

$$(1 + \alpha)^{k+1} > (1 + k\alpha)(1 + \alpha), \tag{3}$$

which can be written in the form

$$(1 + \alpha)^{k+1} > 1 + (k + 1)\alpha + k\alpha^2.$$

Dropping the positive term $k\alpha^2$ from the right side of the last inequality, we obtain inequality (2).

PROBLEM 48.   Prove that for every natural number $n > 1$,

$$\frac{1}{\sqrt{1}} + \frac{1}{\sqrt{2}} + \cdots + \frac{1}{\sqrt{n}} > \sqrt{n}.$$

PROBLEM 49.   Prove that for every natural number $n > 1$,

$$\frac{4^n}{n + 1} < \frac{(2n)!}{(n!)^2}.$$

PROBLEM 50.   Prove that

$$2^{n-1}(a^n + b^n) > (a + b)^n, \tag{1}$$

where $a + b > 0$, $a \neq b$, and n is a natural number greater than 1.

SOLUTION.   1)   For $n = 2$ inequality (1) takes the form

$$2(a^2 + b^2) > (a + b)^2. \tag{2}$$

Since $a \neq b$, we have the inequality

$$(a - b)^2 > 0. \tag{3}$$

By adding $(a + b)^2$ to each side of inequality (3), we obtain inequality (2). This proves that inequality (1) holds for $n = 2$.

2) Let us assume that inequality (1) holds for some natural number $n = k$; that is,

$$2^{k-1}(a^k + b^k) > (a + b)^k. \tag{4}$$

Let us prove that inequality (1) must then hold for $n = k + 1$; that is,

$$2^k(a^{k+1} + b^{k+1}) > (a + b)^{k+1}. \tag{5}$$

Let us multiply both sides of inequality (4) by $a + b$. Since, by our original assumption, $a + b > 0$, we obtain the following inequality:

$$2^{k-1}(a^k + b^k)(a + b) > (a + b)^{k+1}. \tag{6}$$

To prove inequality (5) it is sufficient to show that

$$2^k(a^{k+1} + b^{k+1}) > 2^{k-1}(a^k + b^k)(a + b), \tag{7}$$

or, equivalently,

$$a^{k+1} + b^{k+1} > a^k b + ab^k. \tag{8}$$

Inequality (8) can be written in the form

$$(a^k - b^k)(a - b) > 0. \tag{9}$$

Suppose that $a > b$. Then, from our assumption that $a > 0$, it follows that $a > |b|$; therefore $a^k > b^k$. Thus, the left side of inequality (9) is the product of two positive numbers. If $a < b$ then, by similar reasoning, we see that $a^k < b^k$. In this case the left side of inequality (9) is the product of two negative numbers. In either case inequality (9) holds.

This proves that if inequality (1) holds for $n = k$, it must hold also for $n = k + 1$.

PROBLEM 51.   Prove that for any $x > 0$, and for any natural number n, the following inequality holds:

$$x^n + x^{n-2} + x^{n-4} + \cdots + \frac{1}{x^{n-4}} + \frac{1}{x^{n-2}} + \frac{1}{x^n} \geq n + 1. \tag{1}$$

SOLUTION.   1a)   For $n = 1$, inequality (1) takes the form

$$x + \frac{1}{x} \geq 2. \tag{2}$$

Inequality (2) follows from the obvious inequality

$$(x - 1)^2 \geq 0.$$

1b)   For $n = 2$, inequality (1) takes the form

$$x^2 + 1 + \frac{1}{x^2} \geq 3. \tag{3}$$

Since inequality (2) holds for any $x > 0$, it holds also if x is replaced by x^2, that is,

$$x^2 + \frac{1}{x^2} \geq 2.$$

Adding 1 to both sides of the last inequality, we obtain inequality (3).

2)   Let us assume that inequality (1) holds for some natural number $n = k$; that is,

$$x^k + x^{k-2} + \cdots + \frac{1}{x^{k-2}} + \frac{1}{x^k} \geq k + 1. \tag{4}$$

Let us show that inequality (1) must then hold also for $n = k + 2$; that is,

$$x^{k+2} + x^k + x^{k-2} + \cdots + \frac{1}{x^{k-2}} + \frac{1}{x^k} + \frac{1}{x^{k+2}} \geq k + 3. \tag{5}$$

By replacing x by x^{k+2} in inequality (2), we obtain

$$x^{k+2} + \frac{1}{x^{k+2}} \geq 2. \tag{6}$$

By adding inequalities (4) and (6) termwise, we obtain inequality (5).

Let us summarize our results. In 1a) and 1b) we proved that inequality (1) holds for $n = 1$ and $n = 2$. In 2) we proved that the validity of inequality (1) for $n = k + 2$ follows from its validity for $n = k$. In other words, 2) allows us to proceed from $n = k$ to $n = k + 2$. The results of 1a) and 2) permit us to assert that inequality (1) holds for all *odd* values of n. Similarly, the results of 1b) and 2) permit us to assert that the inequality holds for all *even* values of n. Hence, we can assert that inequality (1) holds for all natural numbers n.

PROBLEM 52. Prove the following theorem: The geometric mean of a finite number of positive numbers is not greater than their arithmetic mean; that is, for any positive numbers $a_1, a_2, \ldots, a_n$,

$$\sqrt[n]{a_1 a_2 \cdots a_n} \leq \frac{a_1 + a_2 + \cdots + a_n}{n}. \tag{1}$$

SOLUTION. 1) For $n = 2$, inequality (1) takes the form

$$\sqrt{a_1 a_2} \leq \frac{a_1 + a_2}{2}. \tag{2}$$

For arbitrary numbers a_1 and a_2, the following inequality holds:

$$(\sqrt{a_1} - \sqrt{a_2})^2 \geq 0.$$

From this inequality it is easy to get inequality (2).

Inequality (2) has a simple geometric interpretation. On a given straight line choose a point A; on this line lay off segments AC and CB having lengths a_1 and a_2, respectively. Now let us construct a circle having the line segment AB as a diameter. Thus, the length of a radius of the circle will be $\frac{a_1 + a_2}{2}$. At the point C let us erect a perpendicular to the line segment AB, and call the point in which this perpendicular intersects the circle D. Then the length of the segment CD is $\sqrt{a_1 a_2}$.

2a) Let us assume that inequality (1) holds for $n = k$, and prove that it then holds also for $n = 2k$.

$$\begin{aligned}
\sqrt[2k]{a_1 a_2 \cdots a_{2k}} &= \sqrt{\sqrt[k]{a_1 a_2 \cdots a_k} \cdot \sqrt[k]{a_{k+1} a_{k+2} \cdots a_{2k}}} \\
&\leq \frac{\sqrt[k]{a_1 a_2 \cdots a_k} + \sqrt[k]{a_{k+1} a_{k+2} \cdots a_{2k}}}{2} \\
&\leq \frac{\dfrac{a_1 + a_2 + \cdots + a_k}{k} + \dfrac{a_{k+1} + a_{k+2} + \cdots + a_{2k}}{k}}{2} \\
&= \frac{a_1 + a_2 + \cdots + a_k + \cdots + a_{2k}}{2k}.
\end{aligned}$$

Since inequality (1) has been verified for $n = 2$, we may assert that it also holds for $n = 4, 8, 16$, etc., that is, for any $n = 2^s$, where s is a natural number.

2b) To prove that inequality (1) holds for all natural numbers n, let us show that from its validity for $n = k$ follows its validity for $n = k - 1$. Let $a_1, a_2, \ldots, a_{k-1}$ be arbitrary positive numbers, and let λ be another, as yet undetermined, positive number. Then

$$\sqrt[k]{a_1 a_2 \cdots a_{k-1} \lambda} \leq \frac{a_1 + a_2 + \cdots + a_{k-1} + \lambda}{k}.$$

Let us choose λ in such a way that

$$\frac{a_1 + a_2 + \cdots + a_{k-1} + \lambda}{k} = \frac{a_1 + a_2 + \cdots + a_{k-1}}{k - 1},$$

that is, let us put

$$\lambda = \frac{a_1 + a_2 + \cdots + a_{k-1}}{k - 1}.$$

We then obtain

$$\sqrt[k]{\frac{a_1 a_2 \cdots a_{k-1}(a_1 + a_2 + \cdots + a_{k-1})}{k - 1}} \leq \frac{a_1 + a_2 + \cdots + a_{k-1}}{k - 1},$$

or

$$\sqrt[k-1]{a_1 a_2 \cdots a_{k-1}} \leq \frac{a_1 + a_2 + \cdots + a_{k-1}}{k - 1}.$$

Now let m be an arbitrary natural number. If $m = 2^s$ for some natural number s, then, according to 2a), inequality (1) holds. If, on the other hand, $m \neq 2^s$, one can find a natural number s such that $m < 2^s$. Then, by virtue of 2a) and 2b), we may assert that the inequality holds also for $n = m$. (This clever proof was given by A. Cauchy, a French mathematician, 1789–1857.)

3. Proofs of Some Theorems of Algebra

8. POLYNOMIALS AND PROGRESSIONS

THEOREM 1. *The square of a polynomial is equal to the sum of the squares of all of its terms, plus the sum of all possible double products of pairs of different terms; or, symbolically,*

$$(a_1 + a_2 + \cdots + a_n)^2$$
$$= a_1{}^2 + a_2{}^2 + \cdots + a_n{}^2$$
$$+ 2(a_1a_2 + a_1a_3 + \cdots + a_{n-1}a_n). \tag{1}$$

Proof. For $n = 2$, formula (1) may be proved by direct multiplication.

Let us assume formula (1) holds for $n = k - 1$, that is,

$$(a_1 + a_2 + \cdots + a_{k-1})^2 = a_1{}^2 + a_2{}^2 + \cdots + a_{k-1}{}^2 + 2S,$$

where S is the sum of all possible products of any two of the numbers $a_1, a_2, \ldots, a_{k-1}$. Let us prove that

$$(a_1 + a_2 + \cdots + a_{k-1} + a_k)^2$$
$$= a_1{}^2 + a_2{}^2 + \cdots + a_{k-1}{}^2 + a_k{}^2 + 2S_1,$$

where S_1 is the sum of all possible products of two of the numbers $a_1, a_2, \ldots, a_{k-1}, a_k$, that is,

$$S_1 = S + (a_1 + a_2 + \cdots + a_{k-1})a_k.$$

Indeed,

$$(a_1 + \cdots + a_{k-1} + a_k)^2 = [(a_1 + \cdots + a_{k-1}) + a_k]^2$$
$$= (a_1 + \cdots + a_{k-1})^2 + 2(a_1 + \cdots + a_{k-1})a_k + a_k{}^2$$
$$= a_1{}^2 + \cdots + a_{k-1}{}^2 + 2S + 2(a_1 + \cdots + a_{k-1})a_k + a_k{}^2$$
$$= a_1{}^2 + a_2{}^2 + \cdots + a_k{}^2 + 2S_1.$$

THEOREM 2. *The nth term a_n of an arithmetic progression is given by*

$$a_n = a_1 + d(n - 1), \tag{1}$$

where a_1 is the first term, and d the common difference.

Proof. Formula (1) holds for $n = 1$.

Let us assume that formula (1) holds for some $n = k$, that is,

$$a_k = a_1 + d(k - 1).$$

Then

$$a_{k+1} = a_k + d = a_1 + d(k - 1) + d = a_1 + dk;$$

that is, formula (1) holds for $n = k + 1$, also.

THEOREM 3. *The nth term a_n of a geometric progression is given by*

$$a_n = a_1 q^{n-1}, \tag{1}$$

where a_1 is the first term, and q the common ratio.

Proof. Formula (1) holds for $n = 1$.

Assume $a_k = a_1 q^{k-1}$.

Then

$$a_{k+1} = a_k q = a_1 q^k.$$

9. PERMUTATIONS, COMBINATIONS, AND THE BINOMIAL THEOREM

THEOREM 4. *The number P_m of permutations of m elements is given by*

$$P_m = m!. \tag{1}$$

Proof. Let us note first of all that $P_1 = 1$, so that formula (1) holds for $m = 1$.

Suppose $P_k = k!$. Let us prove that

$$P_{k+1} = (k + 1)!.$$

From among the given $k + 1$ elements $a_1, a_2, \ldots, a_k, a_{k+1}$, let us take just the first k elements and consider all possible permutations among them. By assumption, there are $k!$ such permutations. Into each of these permutations let us introduce the element a_{k+1}, placing it successively before the first, before the second, . . . , before the kth, and after the kth element. In this way a single permutation of k elements gives rise to $k + 1$ permutations of $k + 1$ elements. In all, therefore, we have $k!\,(k + 1) = (k + 1)!$ permutations of $k + 1$ elements.

Two points remain to be clarified:

1) Among the $(k + 1)!$ permutations are there any two which are the same?

2) Have we accounted for all the permutations possible?

1) Let us assume that among the $(k + 1)!$ permutations there are two which are the same. Let us call them p_1 and p_2. If in the permutation p_1 element a_{k+1} stands in the sth place from the left, it must also stand in the sth place from the left in the permutation p_2. Now let us remove the element a_{k+1} from both p_1 and p_2, thereby obtaining two identical permutations $\bar{p}_1$ and $\bar{p}_2$ of k elements.

Thus, to obtain the permutations p_1 and p_2, we must introduce the element a_{k+1} at the same place in each of the permutations $\bar{p}_1$ and $\bar{p}_2$. But $\bar{p}_1$ and $\bar{p}_2$ are the same permutation; therefore p_1 and p_2 are the same permutation, contrary to our assumption.

2) Let us assume that there is a permutation p of the $k + 1$ elements which we failed to obtain. Let the element a_{k+1} stand in the sth place from the left in the permutation p. Upon removing a_{k+1} from p, we obtain a permutation $\bar{p}$ of the first k elements. This means that to obtain the permutation p it is sufficient to introduce the element a_{k+1} into the permutation $\bar{p}$ at the sth place from the left.

The permutation $\bar{p}$ must have been one of the $k!$ permutations of the elements $a_1, a_2, \ldots, a_k$, for we assumed that this collection of $k!$ permutations included *all possible* permutations of those k elements. Moreover, in constructing the permutations of $k + 1$ elements we introduced a_{k+1} into the first, the second, $\ldots$, and the $(k + 1)$st places from the left in each of these $k!$ permutations. In particular, therefore, we must have introduced a_{k+1} into the sth place of the permutation $\bar{p}$.

Thus, *our permutations are all different from one another, and every possible permutation of the $k + 1$ elements is accounted for.* Hence,

$$P_{k+1} = (k + 1)!.$$

THEOREM 5. *Given m elements, the number of permutations of n of those elements is*

$$A_m^n = m(m - 1) \cdots (m - n + 1). \tag{1}$$

Proof. First of all, let us note that formula (1) holds for $n = 1$, since $A_m^1 = m$.

Let us assume that

$$A_m^k = m(m - 1) \cdots (m - k + 1),$$

where $k < m$. Let us prove that

$$A_m^{k+1} = m(m - 1) \cdots (m - k).$$

To obtain all permutations with $k + 1$ elements taken from the given m elements, it is sufficient to add at the end of each of the permutations of k elements one of the $m - k$ remaining elements. One can readily see that *all* of the permutations of order $k + 1$ which are obtained in this manner *are different from one another,* and that *all possible permutations of the kind under consideration are accounted for.* Hence,

$$A_m^{k+1} = A_m^k(m - k) = m(m - 1) \cdots (m - k).$$

THEOREM 6. *Given a set of m elements, the number of sets of n elements ($n \le m$) which can be formed from these m elements (or, as we say, the number of combinations of m things taken n at a time) is*

$$C_m^n = \frac{m(m - 1) \cdots (m - n + 1)}{1 \cdot 2 \cdot \,\cdots\, \cdot n}. \tag{1}$$

Proof. First of all, let us note that $C_m^1 = m$, so that formula (1) holds for $n = 1$.

Let us assume that

$$C_m^k = \frac{m(m - 1) \cdots (m - k + 1)}{1 \cdot 2 \cdot \,\cdots\, \cdot k}.$$

Let us prove that

$$C_m^{k+1} = \frac{m(m - 1) \cdots (m - k + 1)(m - k)}{1 \cdot 2 \cdot \,\cdots\, \cdot k(k + 1)}.$$

To obtain all possible sets with $k + 1$ elements from a set of m elements, let us write out all sets with k elements from among the m elements, and adjoin to each of them as the $(k + 1)$st element one of the remaining $m - k$ elements. It is clear that in this manner one obtains all sets with $k + 1$ elements from among the m elements, and that each of them is obtained $k + 1$ times. In fact, the set consisting of the elements $a_1, a_2, \ldots, a_k, a_{k+1}$ is obtained by adjoining to the set whose elements are $a_2, a_3, \ldots, a_k, a_{k+1}$ the element a_1; also one may obtain the former set by adjoining to the set whose elements are $a_1, a_3, \ldots, a_k, a_{k+1}$ the element a_2, etc.; and, finally, by adjoining the element a_{k+1} to the set whose elements are $a_1, a_2, \ldots, a_k$. Thus,

$$C_m^{k+1} = C_m^k \frac{m - k}{k + 1} = \frac{m(m - 1) \cdots (m - k)}{1 \cdot 2 \cdot \cdots \cdot k(k + 1)}.$$

THEOREM 7. *For any two numbers a and b, and for every natural number n,*

$$(a + b)^n = a^n + C_n^1 a^{n-1} b + \cdots + C_n^s a^{n-s} b^s + \cdots$$
$$+ C_n^{n-1} ab^{n-1} + b^n \tag{1}$$

(Newton's binomial theorem).

Proof. For $n = 1$ the formula gives $a + b = a + b$; that is, formula (1) holds for $n = 1$.

Let us assume that

$$(a + b)^k = a^k + C_k^1 a^{k-1} b + C_k^2 a^{k-2} b^2 + \cdots + b^k.$$

Then

$$\begin{aligned}
(a + b)^{k+1} &= (a + b)^k (a + b) \\
&= (a^k + C_k^1 a^{k-1} b + \cdots + b^k)(a + b) \\
&= a^{k+1} + (1 + C_k^1) a^k b + (C_k^1 + C_k^2) a^{k-1} b^2 + \cdots \\
&\quad + (C_k^s + C_k^{s+1}) a^{k-s} b^{s+1} + \cdots + b^{k+1}.
\end{aligned}$$

In view of the fact that $C_k^s + C_k^{s+1} = C_{k+1}^{s+1}$, we obtain

$$\begin{aligned}
(a + b)^{k+1} &= a^{k+1} + C_{k+1}^1 a^k b + C_{k+1}^2 a^{k-1} b^2 + \cdots \\
&\quad + C_{k+1}^{s+1} a^{k-s} b^{s+1} + \cdots + b^{k+1}.
\end{aligned}$$

Solutions of Exercises
in Chapter 2

PROBLEM 3.   Hypothesis: $u_n = 3n - 2$.

1) The hypothesis holds for $n = 1$.

2) Assume $u_k = 3k - 2$.

Then $u_{k+1} = u_k + 3 = 3k - 2 + 3 = 3(k + 1) - 2$.

PROBLEM 4.   Hypothesis: $S_n = 2^n - 1$.

1) The hypothesis holds for $n = 1$.

2) Assume $S_k = 2^k - 1$.

Then $S_{k+1} = S_k + 2^k = 2^{k+1} - 1$.

PROBLEM 6.   1)   The formula holds for $n = 1$.

2)   Assume $1^2 + 2^2 + 3^2 + \cdots + k^2 = \dfrac{k(k + 1)(2k + 1)}{6}$.

Then $1^2 + 2^2 + 3^2 + \cdots + k^2 + (k + 1)^2$

$= \dfrac{k(k + 1)(2k + 1)}{6} + (k + 1)^2 = \dfrac{(k + 1)(k + 2)(2k + 3)}{6}$.

PROBLEM 8.   1)   The formula holds for $n = 1$.

2)   Assume $1^2 + 3^2 + 5^2 + \cdots + (2k - 1)^2 = \dfrac{k(2k - 1)(2k + 1)}{3}$.

Then $1^2 + 3^2 + \cdots + (2k - 1)^2 + (2k + 1)^2$

$= \dfrac{k(2k - 1)(2k + 1)}{3} + (2k + 1)^2 = \dfrac{(k + 1)(2k + 1)(2k + 3)}{3}$.

PROBLEM 9.   1)   The formula holds for $n = 1$.

2)   Assume $1^3 + 2^3 + \cdots + k^3 = \left[\dfrac{k(k + 1)}{2} \right]^2$.

Then $1^3 + 2^3 + \cdots + k^3 + (k + 1)^3 = \dfrac{k^2(k + 1)^2}{4} + (k + 1)^3$

$= \left[\dfrac{(k + 1)(k + 2)}{2} \right]^2$.

PROBLEM 10. 1) The formula holds for $n = 1$.

2) Assume $1 + x + x^2 + \cdots + x^k = \dfrac{x^{k+1} - 1}{x - 1}$.

Then $1 + x + x^2 + \cdots + x^k + x^{k+1} = \dfrac{x^{k+1} - 1}{x - 1} + x^{k+1} = \dfrac{x^{k+2} - 1}{x - 1}$.

PROBLEM 11. 1) The formula holds for $n = 1$.

2) Assume $1 \cdot 2 + 2 \cdot 3 + \cdots + k(k + 1) = \dfrac{k(k + 1)(k + 2)}{3}$.

Then $1 \cdot 2 + 2 \cdot 3 + \cdots + k(k + 1) + (k + 1)(k + 2)$

$$= \frac{k(k + 1)(k + 2)}{3} + (k + 1)(k + 2)$$

$$= (k + 1)(k + 2)\left(\frac{k}{3} + 1\right) = \frac{(k + 1)(k + 2)(k + 3)}{3}.$$

PROBLEM 12. 1) The formula holds for $n = 1$.

2) Assume

$$1 \cdot 2 \cdot 3 + 2 \cdot 3 \cdot 4 + \cdots + k(k + 1)(k + 2) = \frac{k(k + 1)(k + 2)(k + 3)}{4}.$$

Then $1 \cdot 2 \cdot 3 + 2 \cdot 3 \cdot 4 + \cdots + k(k + 1)(k + 2) + (k + 1)(k + 2)(k + 3)$

$$= \frac{k(k + 1)(k + 2)(k + 3)}{4} + (k + 1)(k + 2)(k + 3)$$

$$= \frac{(k + 1)(k + 2)(k + 3)(k + 4)}{4}$$

PROBLEM 13. 1) The formula holds for $n = 1$.

2) Assume $\dfrac{1}{1 \cdot 3} + \dfrac{1}{3 \cdot 5} + \cdots + \dfrac{1}{(2k - 1)(2k + 1)} = \dfrac{k}{2k + 1}$.

Then $\dfrac{1}{1 \cdot 3} + \dfrac{1}{3 \cdot 5} + \cdots + \dfrac{1}{(2k - 1)(2k + 1)} + \dfrac{1}{(2k + 1)(2k + 3)}$

$$= \frac{k}{2k + 1} + \frac{1}{(2k + 1)(2k + 3)} = \frac{k + 1}{2k + 3}.$$

PROBLEM 14. 1) The formula holds for $n = 1$.

2) Assume $\dfrac{1^2}{1 \cdot 3} + \dfrac{2^2}{3 \cdot 5} + \cdots + \dfrac{k^2}{(2k - 1)(2k + 1)} = \dfrac{k(k + 1)}{2(2k + 1)}$.

Then $\dfrac{1^2}{1 \cdot 3} + \dfrac{2^2}{3 \cdot 5} + \cdots + \dfrac{k^2}{(2k - 1)(2k + 1)} + \dfrac{(k + 1)^2}{(2k + 1)(2k + 3)}$

$$= \frac{k(k + 1)}{2(2k + 1)} + \frac{(k + 1)^2}{(2k + 1)(2k + 3)} = (k + 1)\frac{k(2k + 3) + 2(k + 1)}{2(2k + 1)(2k + 3)}$$

$$= \frac{(k + 1)(2k^2 + 5k + 2)}{2(2k + 1)(2k + 3)} = \frac{(k + 1)(2k + 1)(k + 2)}{2(2k + 1)(2k + 3)} = \frac{(k + 1)(k + 2)}{2(2k + 3)}.$$

PROBLEM 15. 1) The formula holds for $n = 1$.

2) Assume $\dfrac{1}{1\cdot4} + \dfrac{1}{4\cdot7} + \cdots + \dfrac{1}{(3k-2)(3k+1)} = \dfrac{k}{3k+1}$.

Then $\dfrac{1}{1\cdot4} + \dfrac{1}{4\cdot7} + \cdots + \dfrac{1}{(3k-2)(3k+1)} + \dfrac{1}{(3k+1)(3k+4)}$

$$= \frac{k}{3k+1} + \frac{1}{(3k+1)(3k+4)} = \frac{k+1}{3k+4}.$$

PROBLEM 16. 1) The formula holds for $n = 1$.

2) Assume $\dfrac{1}{1\cdot5} + \dfrac{1}{5\cdot9} + \cdots + \dfrac{1}{(4k-3)(4k+1)} = \dfrac{k}{4k+1}$.

Then $\dfrac{1}{1\cdot5} + \dfrac{1}{5\cdot9} + \cdots + \dfrac{1}{(4k-3)(4k+1)} + \dfrac{1}{(4k+1)(4k+5)}$

$$= \frac{k}{4k+1} + \frac{1}{(4k+1)(4k+5)} = \frac{k+1}{4k+5}.$$

PROBLEM 17. 1) The formula holds for $n = 1$.

2) Assume

$$\frac{1}{a(a+1)} + \frac{1}{(a+1)(a+2)} + \cdots + \frac{1}{(a+k-1)(a+k)} = \frac{k}{a(a+k)}.$$

Then

$$\frac{1}{a(a+1)} + \frac{1}{(a+1)(a+2)} + \cdots + \frac{1}{(a+k-1)(a+k)}$$
$$+ \frac{1}{(a+k)(a+k+1)}$$
$$= \frac{k}{a(a+k)} + \frac{1}{(a+k)(a+k+1)} = \frac{k+1}{a(a+k+1)}.$$

PROBLEM 19. 1) The formula holds for $n = 1$ and for $n = 2$.

2) Assume $u_{k-2} = \dfrac{\alpha^{k-1} - \beta^{k-1}}{\alpha - \beta}, \quad u_{k-1} = \dfrac{\alpha^k - \beta^k}{\alpha - \beta}$.

Then $u_k = (\alpha + \beta)\dfrac{\alpha^k - \beta^k}{\alpha - \beta} - \alpha\beta\dfrac{\alpha^{k-1} - \beta^{k-1}}{\alpha - \beta} = \dfrac{\alpha^{k+1} - \beta^{k+1}}{\alpha - \beta}$.

PROBLEM 21. 1) For $n = 0$ the formula takes the form

$$\frac{1}{1+x} = \frac{1}{x-1} + \frac{2}{1-x^2}.$$

Thus we see that the formula holds in this case.

2) Assume

$$\frac{1}{1+x} + \frac{2}{1+x^2} + \frac{4}{1+x^4} + \cdots + \frac{2^k}{1+x^{2^k}} = \frac{1}{x-1} + \frac{2^{k+1}}{1-x^{2^{k+1}}}.$$

Then $\quad \dfrac{1}{1+x} + \dfrac{2}{1+x^2} + \dfrac{4}{1+x^4} + \cdots + \dfrac{2^k}{1+x^{2^k}} + \dfrac{2^{k+1}}{1+x^{2^{k+1}}}$

$$= \dfrac{1}{x-1} + \dfrac{2^{k+1}}{1-x^{2^{k+1}}} + \dfrac{2^{k+1}}{1+x^{2^{k+1}}} = \dfrac{1}{x-1} + \dfrac{2^{k+2}}{1-x^{2^{k+2}}}.$$

(See the Second Remark of section 4 in Chapter 1.)

PROBLEM 23.   For $n = 1$ we have

$$1 - \frac{x}{1!} = -\frac{x-1}{1}.$$

For $n = 2$ we have

$$1 - \frac{x}{1!} + \frac{x(x-1)}{2!} = -\frac{x-1}{1} + \frac{x(x-1)}{2} = \frac{(x-1)(x-2)}{2!}.$$

For $n = 3$ we have

$$1 - \frac{x}{1!} + \frac{x(x-1)}{2!} - \frac{x(x-1)(x-2)}{3!}$$

$$= \frac{(x-1)(x-2)}{2} - \frac{x(x-1)(x-2)}{6} = -\frac{(x-1)(x-2)(x-3)}{3!}.$$

This suggests the hypothesis that

$$1 - \frac{x}{1!} + \frac{x(x-1)}{2!} - \cdots + (-1)^n \frac{x(x-1)\cdots(x-n+1)}{n!}$$

$$= (-1)^n \frac{(x-1)(x-2)\cdots(x-n)}{n!}.$$

1)   The hypothesis holds for $n = 1$.

2)   Assume

$$1 - \frac{x}{1!} + \frac{x(x-1)}{2!} - \cdots + (-1)^k \frac{x(x-1)\cdots(x-k+1)}{k!}$$

$$= (-1)^k \frac{(x-1)(x-2)\cdots(x-k)}{k!}.$$

Then

$$1 - \frac{x}{1!} + \frac{x(x-1)}{2!} - \cdots + (-1)^k \frac{x(x-1)\cdots(x-k+1)}{k!}$$

$$+ (-1)^{k+1} \frac{x(x-1)\cdots(x-k)}{(k+1)!}$$

$$= (-1)^k \frac{(x-1)(x-2)\cdots(x-k)}{k!} + (-1)^{k+1} \frac{x(x-1)\cdots(x-k)}{(k+1)!}$$

$$= (-1)^{k+1} \frac{(x-1)(x-2)\cdots(x-k)}{k!} \left[\frac{x}{k+1} - 1 \right]$$

$$= (-1)^{k+1} \frac{(x-1)(x-2)\cdots(x-k)(x-k-1)}{(k+1)!}.$$

Problem 26. 1) The formula holds for $n = 0$.

2) Let us assume that the formula holds for $n = k$, that is, that

$$A_k = 11^{k+2} + 12^{2k+1}$$

is divisible by 133. Then

$$\begin{aligned}
A_{k+1} &= 11^{k+3} + 12^{2(k+1)+1} = 11^{k+3} + 12^{2k+3}\\
&= 11 \cdot 11^{k+2} + 144 \cdot 12^{2k+1}\\
&= 11 \cdot 11^{k+2} + 133 \cdot 12^{2k+1} + 11 \cdot 12^{2k+1}\\
&= 11 \cdot (11^{k+2} + 12^{2k+1}) + 133 \cdot 12^{2k+1}\\
&= 11A_k + 133 \cdot 12^{2k+1}.
\end{aligned}$$

We have shown that A_{k+1} is the sum of two numbers each of which is divisible by 133; consequently, A_{k+1} is divisible by 133.

Problem 28. The statement is true in the case where $n = 1$, since a single straight line divides the plane into 2 parts.

Let us assume that k different straight lines which pass through a given point divide the plane into $2k$ parts. Then if we have a $(k + 1)$st line also passing through the given point, it will divide two of the original parts in two. Hence the plane will be divided into $2(k + 1)$ parts.

Problem 29. 1) The straight line AB divides the plane P into two half-planes, P_1 and P_2. Let us color P_1 white and P_2 black, to satisfy the requirements of the problem. Thus, we see that the statement is true for $n = 1$.

2) Assume the statement holds for $n = k$, and suppose that plane P has been colored in accordance with the requirements of the problem. Let the $(k + 1)$st straight line CD divide the plane into two half-planes, Q_1 and Q_2. Throughout Q_1 let us leave the coloring unchanged; throughout Q_2, however, let us replace white by black and black by white.

Now let O_1 and O_2 be any two neighboring regions of the figure after the line CD is drawn. There are two possibilities to consider:

a) O_1 and O_2 lie on opposite sides of CD,

b) O_1 and O_2 lie on the same side of CD.

In the first case, O_1 and O_2 must have formed a single region before the line CD was drawn, and hence they had the same color. After CD was drawn, the one which happened to lie in Q_1 retained its coloring with no change, while the one which happened to lie in Q_2 had its coloring changed.

In the second case, before CD was drawn, O_1 and O_2 were part of two different neighboring regions whose common boundary was one of the original k straight lines; consequently, they must have been colored differently. If, after CD was drawn, both came to lie in Q_1, the coloring of each remained unchanged. If, on the other hand, they both came to lie in Q_2, the coloring of each was changed. In either case, O_1 and O_2 have different colors.

PROBLEM 34.   1)   The formula holds for $n = 1$, since

$$\frac{\sin \frac{3x}{2}}{2 \sin \frac{x}{2}} = \frac{\sin \frac{x}{2} + \left(\sin \frac{3x}{2} - \sin \frac{x}{2} \right)}{2 \sin \frac{x}{2}} = \frac{1}{2} + \cos x.$$

2)   Assume $\dfrac{1}{2} + \cos x + \cos 2x + \cdots + \cos kx = \dfrac{\sin \frac{2k+1}{2} x}{2 \sin \frac{x}{2}}.$

Then

$$\frac{1}{2} + \cos x + \cos 2x + \cdots + \cos kx + \cos (k+1)x$$

$$= \frac{\sin \frac{2k+1}{2} x}{2 \sin \frac{x}{2}} + \cos (k+1)x = \frac{\sin \frac{2k+1}{2} x + 2 \sin \frac{x}{2} \cos (k+1)x}{2 \sin \frac{x}{2}}$$

$$= \frac{\sin \frac{2k+1}{2} x + \left(\sin \frac{2k+3}{2} x - \sin \frac{2k+1}{2} x \right)}{2 \sin \frac{x}{2}} = \frac{\sin \frac{2k+3}{2} x}{2 \sin \frac{x}{2}}.$$

PROBLEM 35.   1)   The formula holds for $n = 1$, since

$$\frac{2 \sin x - \sin 2x}{4 \sin^2 \frac{x}{2}} = \frac{2 \sin x (1 - \cos x)}{4 \sin^2 \frac{x}{2}}$$

$$= \frac{2 \sin x \left(1 - \cos 2 \left(\frac{x}{2} \right) \right)}{4 \sin^2 \frac{x}{2}} = \frac{2 \sin x \left(1 - \cos^2 \frac{x}{2} + \sin^2 \frac{x}{2} \right)}{4 \sin^2 \frac{x}{2}} = \sin x.$$

2)   Assume

$$\sin x + 2 \sin 2x + \cdots + k \sin kx = \frac{(k+1) \sin kx - k \sin (k+1)x}{4 \sin^2 \frac{x}{2}}.$$

Then

$$\sin x + 2 \sin 2x + \cdots + k \sin kx + (k+1) \sin (k+1)x$$

$$= \frac{(k+1) \sin kx - k \sin (k+1)x}{4 \sin^2 \frac{x}{2}} + (k+1) \sin (k+1)x$$

$$= \frac{(k+1) \sin kx - k \sin (k+1)x + 2(k+1) \sin (k+1)x (1 - \cos x)}{4 \sin^2 \frac{x}{2}}$$

$$= \frac{(k+2)\sin(k+1)x + (k+1)\sin kx}{4\sin^2\frac{x}{2}} - \frac{2(k+1)\cos x \sin(k+1)x}{4\sin^2\frac{x}{2}}$$

$$= \frac{(k+2)\sin(k+1)x + (k+1)\sin kx}{4\sin^2\frac{x}{2}} - \frac{(k+1)[\sin(k+2)x + \sin kx]}{4\sin^2\frac{x}{2}}$$

$$= \frac{(k+2)\sin(k+1)x - (k+1)\sin(k+2)x}{4\sin^2\frac{x}{2}}.$$

PROBLEM 36. 1) The formula holds for $n = 1$, since

$$\frac{2\cos x - \cos 2x - 1}{4\sin^2\frac{x}{2}} = \frac{2\cos x - 2\cos^2 x}{4\sin^2\frac{x}{2}}$$

$$= \frac{\cos x (1 - \cos x)}{2\sin^2\frac{x}{2}}$$

$$= \cos x.$$

2) Assume

$$\cos x + 2\cos 2x + \cdots + k\cos kx$$
$$= \frac{(k+1)\cos kx - k\cos(k+1)x - 1}{4\sin^2\frac{x}{2}}.$$

Then

$$\cos x + 2\cos 2x + \cdots + k\cos kx + (k+1)\cos(k+1)x$$

$$= \frac{(k+1)\cos kx - k\cos(k+1)x - 1}{4\sin^2\frac{x}{2}} + (k+1)\cos(k+1)x$$

$$= \frac{(k+1)\cos kx - k\cos(k+1)x - 1}{\sin^2\frac{x}{2}} + \frac{2(k+1)\cos(k+1)x(1 - \cos x)}{4\sin^2\frac{x}{2}}$$

$$= \frac{(k+2)\cos(k+1)x + (k+1)\cos kx}{4\sin^2\frac{x}{2}} - \frac{2(k+1)\cos x \cos(k+1)x + 1}{4\sin^2\frac{x}{2}}$$

$$= \frac{(k+2)\cos(k+1)x + (k+1)\cos kx}{4\sin^2\frac{x}{2}} - \frac{(k+1)[\cos(k+2)x + \cos kx] + 1}{4\sin^2\frac{x}{2}}$$

$$= \frac{(k+2)\cos(k+1)x - (k+1)\cos(k+2)x - 1}{4\sin^2\frac{x}{2}}.$$

PROBLEM 37. 1) The formula holds for $n = 1$, since

$$\frac{1}{2} \cot \frac{x}{2} - \cot x = \frac{1}{2} \cot \frac{x}{2} - \frac{1 - \tan^2 \frac{x}{2}}{2 \tan \frac{x}{2}} = \frac{\tan^2 \frac{x}{2}}{2 \tan \frac{x}{2}} = \frac{1}{2} \tan \frac{x}{2}.$$

2) Assume

$$\frac{1}{2} \tan \frac{x}{2} + \frac{1}{2^2} \tan \frac{x}{2^2} + \cdots + \frac{1}{2^k} \tan \frac{x}{2^k} = \frac{1}{2^k} \cot \frac{x}{2^k} - \cot x.$$

Then $\frac{1}{2} \tan \frac{x}{2} + \frac{1}{2^2} \tan \frac{x}{2^2} + \cdots + \frac{1}{2^k} \tan \frac{x}{2^k} + \frac{1}{2^{k+1}} \tan \frac{x}{2^{k+1}}$

$$= \frac{1}{2^k} \cot \frac{x}{2^k} - \cot x + \frac{1}{2^{k+1}} \tan \frac{x}{2^{k+1}}$$

$$= \frac{1}{2^{k+1}} \frac{\cot^2 \frac{x}{2^{k+1}} - 1}{\cot \frac{x}{2^{k+1}}} + \frac{1}{2^{k+1} \cot \frac{x}{2^{k+1}}} - \cot x = \frac{1}{2^{k+1}} \cot \frac{x}{2^{k+1}} - \cot x.$$

PROBLEM 38. 1) Since

$$\tan (\text{arc tan } 2 - \text{arc tan } 1) = \frac{2 - 1}{1 + 2} = \frac{1}{3},$$

it follows that

$$\text{arc tan } 2 - \text{arc tan } 1 = \text{arc tan } \frac{1}{3} = \text{arc cot } 3.$$

Hence, the formula holds for $n = 1$.

2) First let us prove that

$$\text{arc cot } (2k + 3) = \text{arc tan } \frac{k + 2}{k + 1} - \text{arc tan } 1. \tag{1}$$

In fact, $\tan \left(\text{arc tan } \frac{k + 2}{k + 1} - \text{arc tan } 1 \right) = \frac{\frac{k + 2}{k + 1} - 1}{1 + \frac{k + 2}{k + 1}} = \frac{1}{2k + 3}.$

Hence,

$$\text{arc tan } \frac{1}{2k + 3} = \text{arc cot } (2k + 3) = \text{arc tan } \frac{k + 2}{k + 1} - \text{arc tan } 1.$$

Let us assume that the formula holds for $n = k$; that is,

$$\text{arc cot } 3 + \text{arc cot } 5 + \cdots + \text{arc cot } (2k + 1)$$

$$= \text{arc tan } 2 + \text{arc tan } \frac{3}{2} + \cdots + \text{arc tan } \frac{k + 1}{k} - k \text{ arc tan } 1. \tag{2}$$

Let us now prove that the formula then holds also for $n = k + 1$; that is,

$$\text{arc cot } 3 + \text{arc cot } 5 + \cdots + \text{arc cot } (2k + 3)$$

$$= \text{arc tan } 2 + \cdots + \text{arc tan } \frac{k + 2}{k + 1} - (k + 1) \text{ arc tan } 1. \tag{3}$$

By adding equalities (1) and (2) termwise we obtain equality (3).

PROBLEM 40. 1) The formula holds for $n = 1$, since

$$\sqrt{3} - i = 2 \left(\cos \frac{\pi}{6} - i \sin \frac{\pi}{6} \right).$$

2) Assume $(\sqrt{3} - i)^k = 2^k \left(\cos \frac{k\pi}{6} - i \sin \frac{k\pi}{6} \right).$

Then $(\sqrt{3} - i)^{k+1} = 2^k \left(\cos \frac{k\pi}{6} - i \sin \frac{k\pi}{6} \right) \cdot 2 \left(\cos \frac{\pi}{6} - i \sin \frac{\pi}{6} \right)$

$$= 2^{k+1} \left[\cos \frac{(k+1)\pi}{6} - i \sin \frac{(k+1)\pi}{6} \right].$$

PROBLEM 42. 1) The formula holds for $n = 1$.

2) Assume that

$$(\cos x + i \sin x)^k = \cos kx + i \sin kx.$$

Then $(\cos x + i \sin x)^{k+1} = (\cos kx + i \sin kx)(\cos x + i \sin x)$

$$= \cos (k+1)x + i \sin (k+1)x.$$

PROBLEM 44. The very last phrase is a fallacy: "This proves the inequality." In fact we have merely shown that the inequality

$$2^n > 2n + 1$$

holds for $n = k + 1$ if it holds for $n = k$, where k is an arbitrary natural number.

This, however, does not prove that the inequality holds for *even a single* value of n, let alone for all values of n, where n is a natural number. In short, the fallacy consists in proving only condition 2, while condition 1 is not considered, so that no basis has been laid for an inductive proof.

PROBLEM 45. One readily observes that 3 is the least natural number for which the inequality $2^n > 2n + 1$ holds.

Taking into account the fact that the validity of the inequality for $n = k$ implies its validity for $n = k + 1$ (Problem 44), it follows that the inequality holds for every natural number $n \geq 3$.

PROBLEM 48. 1) The formula holds for $n = 2$, since

$$1 + \frac{1}{\sqrt{2}} > \sqrt{2}.$$

2) Assuming that $\dfrac{1}{\sqrt{1}} + \dfrac{1}{\sqrt{2}} + \cdots + \dfrac{1}{\sqrt{k}} > \sqrt{k},$ \hfill (1)

let us prove that

$$\frac{1}{\sqrt{1}} + \frac{1}{\sqrt{2}} + \cdots + \frac{1}{\sqrt{k}} + \frac{1}{\sqrt{k+1}} > \sqrt{k+1}. \tag{2}$$

For any $k \geq 0$ the inequality

$$\frac{1}{\sqrt{k+1}} > \sqrt{k+1} - \sqrt{k} \tag{3}$$

holds. In fact, inequality (3) is obtained from the inequality

$$1 + \sqrt{\frac{k}{k+1}} > 1$$

by multiplying both sides by $\sqrt{k+1} - \sqrt{k}$. Adding inequalities (1) and (3) termwise, we obtain inequality (2).

PROBLEM 49. 1) The inequality holds for $n = 2$, since $\frac{16}{3} < 6$.

2) Assume
$$\frac{4^k}{k+1} < \frac{(2k)!}{(k!)^2},$$

where $k \geq 2$. It is easy to prove that for $k > 0$

$$\frac{4(k+1)}{k+2} < \frac{(2k+1)(2k+2)}{(k+1)^2}.$$

Hence,
$$\frac{4^k}{k+1} \cdot \frac{4(k+1)}{k+2} < \frac{(2k)!}{(k!)^2} \cdot \frac{(2k+1)(2k+2)}{(k+1)^2},$$

that is,
$$\frac{4^{k+1}}{k+2} < \frac{(2k+2)!}{[(k+1)!]^2}.$$

INDUCTION IN
GEOMETRY

L. I. Golovina and I. M. Yaglom

PREFACE TO THE AMERICAN EDITION

THIS BOOKLET is devoted to various applications of the method of mathematical induction to the solution of geometric problems, some of which are rather intricate. The booklet contains 37 examples whose solutions are given in detail and 40 problems for which only brief hints are given. Some of these problems provide a real challenge to test the mathematical mettle of the reader.

For most sections of the text the background of high school algebra and plane geometry will be sufficient. For Chapter 6 some knowledge of solid geometry is necessary. Occasionally formulas from trigonometry are used.

The chapters are very nearly independent of one another, and so it is possible for the reader to omit any chapter for which he lacks the necessary preparation. For more background on mathematical induction the reader may wish to consult another booklet in this series, *The Method of Mathematical Induction* by I. S. Sominskii.

CONTENTS

Introduction: What Is the Method of Mathematical Induction?

1. FAULTY REASONING "BY ANALOGY"

Induction is any reasoning in which one goes from a particular assertion to a general one, the correctness of which is deducible from the correctness of the particular assertion. The *method of mathematical induction* is a special method of mathematical proof, which makes it possible, on the basis of particular observations, to draw conclusions with respect to corresponding general laws. The idea of this method can be explained most simply by examples. We shall begin, therefore, by examining the following example:

EXAMPLE 1. Find the sum of the first n odd numbers

$$1 + 3 + 5 + \cdots + (2n - 1).$$

SOLUTION. If we denote this sum by $S(n)$, and put $n = 1, 2, 3, 4, 5$, then we shall have

$$
\begin{aligned}
S(1) &= 1, \\
S(2) &= 1 + 3 = 4, \\
S(3) &= 1 + 3 + 5 = 9, \\
S(4) &= 1 + 3 + 5 + 7 = 16, \\
S(5) &= 1 + 3 + 5 + 7 + 9 = 25.
\end{aligned}
$$

We notice that for $n = 1, 2, 3, 4, 5$, the sum of the first n odd numbers is equal to n^2. From this fact may we jump to the conclusion that this holds true for any n? No, such a conclusion "by analogy" may turn out to be erroneous.

1

Let us give several examples where a conclusion "by analogy" is indeed incorrect.

First, we consider numbers of the form $2^{2^n} + 1$. For $n = 0, 1, 2, 3$, and 4, the numbers $2^{2^0} + 1 = 3$, $2^{2^1} + 1 = 5$, $2^{2^2} + 1 = 17$, $2^{2^3} + 1 = 257$, $2^{2^4} + 1 = 65{,}537$ are prime numbers. Pierre Fermat, an outstanding French mathematician of the seventeenth century, conjectured that *all* numbers of this form are prime numbers. However, in the eighteenth century another great scholar, Leonhard Euler,[1] found that

$$2^{2^5} + 1 = 4{,}294{,}967{,}297 = 641 \cdot 6{,}700{,}417, \text{ a composite number.}$$

Here is another example of the same type. G. W. Leibniz, the celebrated German mathematician of the seventeenth century, one of the creators of so-called "higher mathematics," proved that for every positive integer n, the number $n^3 - n$ is divisible by 3, the number $n^5 - n$ is divisible by 5, the number $n^7 - n$ is divisible by 7. On this basis he conjectured that for every odd k and for any natural number n, the number $n^k - n$ is divisible by k, but he himself soon discovered that $2^9 - 2 = 510$ is *not* divisible by 9.

The noted Soviet mathematician D. A. Grave once fell into the same type of error. He conjectured that for all prime numbers p, the number $2^{p-1} - 1$ is not divisible by p^2. This conjecture was confirmed by direct examination for all prime numbers p less than one thousand. However, it was soon established that $2^{1092} - 1$ is divisible by 1093 (1093 is a prime number); that is, Grave's conjecture proved to be false.

Let us give one more very convincing example. In the expression $991n^2 + 1$, substituting for n successively the numbers 1, 2, 3, . . . , we never obtain a number which is a perfect square, even if we dedicate days or years to this computation. However, if we conclude from this that *all* numbers of this form are not perfect squares, then we shall be in error, for actually it turns out that numbers of the form $991n^2 + 1$ do include squares, but the smallest value of n for which the number $991n^2 + 1$ is a perfect square is very large. Here is that number:

$$n = 12{,}055{,}735{,}790{,}331{,}359{,}447{,}442{,}538{,}767.$$

---

[1] Leonhard Euler (1707–1783) was a Swiss mathematician who at the age of 20 was appointed to the chair of mathematics at the St. Petersburg Academy.

All of these examples should serve as a warning to the reader not to draw unfounded conclusions by analogy.

2. THE METHOD OF MATHEMATICAL INDUCTION

Let us return now to the problem of computing the sum of the first n odd numbers in Example 1. From the preceding discussion it is clear that the formula

$$S(n) = n^2 \tag{1}$$

cannot be regarded as proved, no matter for how many individual values of n we may have verified it, for there is always the danger that the formula is not valid for some value of n that we have not considered. In order to make sure that formula (1) is valid for all n, it is necessary to prove that no matter how far we move along the sequence of natural numbers, we can never progress from values of n for which formula (1) is still true, to values of n for which the formula is no longer true.

Thus, *let us suppose that for some number n our formula is true, and try to prove that then it is true also for the number n + 1.*

Thus, we assume that

$$S(n) = 1 + 3 + 5 + \ldots + (2n - 1) = n^2;$$

let us calculate

$$S(n + 1) = 1 + 3 + 5 + \ldots + (2n - 1) + (2n + 1).$$

On the strength of our supposition, the sum of the first n terms on the right-hand side of the last equality is equal to n^2; consequently,

$$S(n + 1) = n^2 + (2n + 1) = (n + 1)^2.$$

Thus, by assuming that the formula $S(n) = n^2$ is valid for some natural number n, we have been able to prove that it is valid also for $n + 1$, the number which follows immediately after n. But earlier we proved that this formula was true for $n = 1, 2, 3, 4, 5$. Consequently, it will be true also for $n = 6$, which follows 5, and then it will be true also for $n = 7$, and for $n = 8$, and for $n = 9$, and so on. Now our formula can be regarded as having been proved for *any* number of terms. It is this method of proof which is called *the method of mathematical induction*.

Thus, a proof by *the method of mathematical induction* consists of the following two parts:

1. *Proof that the assertion made is valid for a small natural number n_0 for which the assertion has a meaning.*[1]

2. *Proof that if this assertion is valid for some natural number $n \geq n_0$, then it is valid also for $n + 1$, the number which follows immediately after n.*

Such a proof will show that the assertion made is true for all natural numbers $n \geq n_0$.

We have already convinced ourselves of the necessity for the second part of the proof by several examples. However, it is not sufficient to prove only the second part, since it may turn out that the assertion is not valid for any value of n. For example, if the assertion is that an arbitrary number n is equal to the number which follows it, that is, that $n = n + 1$, then adding 1 to each side of this equality, we obtain $n + 1 = n + 2$; that is, the number $n + 1$ is also equal to the number which follows it. From this, of course, it does not follow at all that the stated assertion is valid for all n; indeed, it is not valid for any number n.

In using the method of mathematical induction, we are not obliged to follow the above scheme exactly. For example, sometimes we have to assume that the assertion in question is valid for, say, *two consecutive* numbers $n - 1$ and n, and prove that the assertion is valid also for $n + 1$; in this case, as the first step of the argument, it is necessary to verify that the assertion is valid for the first *two* values of n, for example, for $n = 1$ and $n = 2$ (see Examples 16, 17, 18 below). Sometimes, as the second step in the argument, the validity of the assertion is proved for some value of n, assuming its validity for all natural numbers k *less than* n (see Examples 7, 8, 9, 15 below).

3. APPLICATIONS

Let us consider several more examples of the application of the method of mathematical induction. The formulas that we obtain will be used later on.

---

[1] It is clear that this value need not always be 1; thus, for example, an assertion about the general properties of a polygon of n sides has meaning only for $n \geq 3$.

EXAMPLE 2. Prove that the sum of the first n natural numbers, which we shall denote by $S_1(n)$, is equal to $n(n+1)/2$, that is,

$$S_1(n) = 1 + 2 + 3 + \cdots + n = \frac{n(n+1)}{2}. \qquad (2)$$

SOLUTION. 1. $S_1(1) = 1 = \dfrac{1(1+1)}{2}$.

2. Let us assume that

$$S_1(n) = 1 + 2 + 3 + \cdots + n = \frac{n(n+1)}{2}.$$

Then

$$S_1(n+1) = 1 + 2 + 3 + \cdots + n + (n+1)$$
$$= \frac{n(n+1)}{2} + (n+1) = \frac{n(n+1) + 2(n+1)}{2}$$
$$= \frac{(n+1)(n+2)}{2} = \frac{(n+1)[(n+1)+1]}{2},$$

which completes the proof.

EXAMPLE 3. Prove that $S_2(n)$, the sum of the squares of the first n natural numbers, is equal to $n(n+1)(2n+1)/6$; that is,

$$S_2(n) = 1^2 + 2^2 + 3^2 + \cdots + n^2 = \frac{n(n+1)(2n+1)}{6}. \qquad (3)$$

SOLUTION. 1. $S_2(1) = 1^2 = \dfrac{1(1+1)(2\cdot 1 + 1)}{6}$.

2. Let us assume that

$$S_2(n) = \frac{n(n+1)(2n+1)}{6}.$$

Then

$$S_2(n+1) = 1^2 + 2^2 + 3^2 + \cdots + n^2 + (n+1)^2$$
$$= \frac{n(n+1)(2n+1)}{6} + (n+1)^2,$$

and finally

$$S_2(n+1) = \frac{(n+1)[(n+1)+1][2(n+1)+1]}{6}.$$

Problem 1. Prove that $S_3(n)$, the sum of the cubes of the first n natural numbers, is equal to $n^2(n + 1)^2/4$; that is,

$$S_3(n) = 1^3 + 2^3 + 3^3 + \cdots + n^3 = \frac{n^2(n + 1)^2}{4}. \qquad (4)$$

EXAMPLE 4. Prove that

$$0 \cdot 1 + 1 \cdot 2 + 2 \cdot 3 + 3 \cdot 4 + \cdots + (n - 1)n = \frac{(n - 1)n(n + 1)}{3}. \qquad (5)$$

SOLUTION. 1. $\quad 1 \cdot 2 = \dfrac{1 \cdot 2 \cdot 3}{3}$.

2. If

$$1 \cdot 2 + 2 \cdot 3 + 3 \cdot 4 + \cdots + (n - 1)n = \frac{(n - 1)n(n + 1)}{3},$$

then

$$1 \cdot 2 + 2 \cdot 3 + 3 \cdot 4 + \cdots + (n - 1)n + n(n + 1)$$
$$= \frac{(n - 1)n(n + 1)}{3} + n(n + 1) = \frac{n(n + 1) \cdot (n + 2)}{3}.$$

Problem 2. Deduce formula (5) from formulas (2) and (3).

Hint. As a preliminary step, show that

$$0 \cdot 1 + 1 \cdot 2 + 2 \cdot 3 + 3 \cdot 4 + \cdots + (n - 1)n$$
$$= (1^2 + 2^2 + 3^2 + \cdots + n^2) - (1 + 2 + 3 + \cdots + n).$$

Since it is so intimately connected with the number concept, the method of mathematical induction has its greatest applications in arithmetic, algebra, and the theory of numbers. Many interesting examples of this type have been collected in the booklet by I. S. Sominskii, to which reference is made in the Preface. But the concept of the whole number is fundamental to mathematics and is not restricted to the theory of numbers, which studies the properties of whole numbers. The method of mathematical induction is used in widely varied fields of mathematics. Unusually beautiful applications of this method are encountered in geometry.

1. Computation by Induction

As in the theory of numbers and algebra, the most natural application of the method of mathematical induction in geometry is to the solution of problems in computation. Let us look at several examples.

4. EXAMPLES PERTAINING TO REGULAR 2^n-GONS

EXAMPLE 5. Compute a_{2^n}, the length of the side of a regular 2^n-gon inscribed in a circle of radius R.

SOLUTION. 1. For $n = 2$, the regular 2^n-gon is a square; its side is

$$a_4 = R\sqrt{2}.$$

Further, $a_{2^{n+1}}$ can be computed from a_{2^n} by means of the following formula:[1]

$$a_{2^{n+1}} = \sqrt{2R^2 - 2R\sqrt{R^2 - \frac{a_{2^n}^2}{4}}}.$$

From this we find for the side of a regular octagon,

$$a_8 = R\sqrt{2 - \sqrt{2}};$$

for a regular 16-gon,

$$a_{16} = R\sqrt{2 - \sqrt{2 + \sqrt{2}}};$$

for a regular 32-gon,

$$a_{32} = R\sqrt{2 - \sqrt{2 + \sqrt{2 + \sqrt{2}}}}.$$

Therefore, we might hypothesize that an inscribed regular 2^n-gon has side of length

$$a_{2^n} = R\sqrt{2 - \underbrace{\sqrt{2 + \sqrt{2 + \cdots + \sqrt{2}}}}_{n - 2 \text{ twos}}}, \tag{6}$$

for any $n \geq 2$.

[1] Derived by application of the Pythagorean theorem.

2. Let us assume that the side of an inscribed regular 2^n-gon is given by formula (6). In this case the formula for $a_{2^{n+1}}$ gives

$$a_{2^{n+1}} = \sqrt{2R^2 - 2R\sqrt{R^2 - R^2 \dfrac{2 - \sqrt{\underbrace{2 + \cdots + \sqrt{2}}_{n-2 \text{ twos}}}}{4}}}$$

$$= R\sqrt{2 - \sqrt{2 + \underbrace{\sqrt{2 + \cdots + \sqrt{2}}}_{n-1 \text{ twos}}}},$$

from which it follows that formula (6) is valid for all n.

From formula (6) it follows that the circumference, $C = 2\pi R$, of a circle of radius R is equal to the limit of the expression for the perimeter of a 2^n-gon,

$$2^n a_{2^n} = 2^n R \sqrt{2 - \underbrace{\sqrt{2 + \cdots + \sqrt{2}}}_{n-2 \text{ twos}}},$$

as n increases without bound. Consequently,

$$2\pi R = \lim_{n \to \infty} 2^n R \sqrt{2 - \underbrace{\sqrt{2 + \cdots + \sqrt{2}}}_{n-2 \text{ twos}}},$$

$$\pi = \lim_{n \to \infty} 2^{n-1} \sqrt{2 - \underbrace{\sqrt{2 + \cdots + \sqrt{2}}}_{n-2 \text{ twos or } [(n-1)-1] \text{ twos}}}$$

$$= \lim_{n \to \infty} 2^n \sqrt{2 - \underbrace{\sqrt{2 + \cdots + \sqrt{2}}}_{n-1 \text{ twos}}}.$$

Problem 3. Using formula (6), prove that π is the limit of the expression

$$\cfrac{2}{\sqrt{\frac{1}{2}} \sqrt{\frac{1}{2}\left(1 + \sqrt{\frac{1}{2}}\right)} \sqrt{\frac{1}{2}\left(1 + \sqrt{\frac{1}{2}\left(1 + \sqrt{\frac{1}{2}}\right)}\right)} \cdots}$$

as the number of factors (square roots) in the denominator increases without bound (VIETA'S FORMULA[1]). The law of formation of the factors is established by the first three terms which are written out.

---

[1] F. Vieta (1540–1603) was a well-known French mathematician and was one of the inventors of our present algebraic symbols.

Hint. Let S_{2^n} denote the area of a regular 2^n-gon inscribed in a circle of radius R, and let h_{2^n} denote the length of its apothem.[1] Then, since

$$h_{2^n} = \sqrt{R^2 - \frac{a_{2^n}^2}{4}},$$

it follows from formula (6) that

$$h_{2^n} = \frac{R}{2}\sqrt{2 + \underbrace{\sqrt{2 + \cdots + \sqrt{2}}}_{n-2\ \text{twos}}}.$$

Thus,

$$S_{2^n} = \frac{1}{2}(2^n a_{2^n})h_{2^n} = 2^{n-1}a_{2^n}h_{2^n} \tag{*}$$

$$= 2^{n-2}R^2\sqrt{2 - \underbrace{\sqrt{2 + \sqrt{2 + \cdots + \sqrt{2}}}}_{n-3\ \text{twos}}}$$

$$= 2^{n-2}Ra_{2^{n-1}}$$

(assuming that $n \geq 3$). Similarly,

$$S_{2^{n+1}} = 2^{n-1}Ra_{2^n}. \tag{**}$$

From (*) and (**) we have

$$\frac{S_{2^n}}{S_{2^{n+1}}} = \frac{2^{n-1}a_{2^n}h_{2^n}}{2^{n-1}Ra_{2^n}} = \frac{h_{2^n}}{R} = \cos\frac{180°}{2^n},$$

from which it follows that

$$\frac{S_4}{S_{2^n}} = \frac{S_4}{S_8}\cdot\frac{S_8}{S_{16}}\cdot \cdots \cdot\frac{S_{2^{n-1}}}{S_{2^n}} = \cos\frac{180°}{4}\cos\frac{180°}{8}\cdots\cos\frac{180°}{2^{n-1}}.$$

Since $S_4 = 2R^2$, and $\lim\limits_{n\to\infty} S_{2^n} = \pi R^2$, then

$$\lim_{n\to\infty}\frac{S_4}{S_{2^n}} = \frac{2R^2}{\pi R^2} = \frac{2}{\pi},$$

and $\dfrac{2}{\pi}$ is the limit of the expression

$$\cos 45° \cos\frac{45°}{2}\cos\frac{45°}{4}\cdots.$$

Then all that remains is to apply the formula

$$\cos\frac{\alpha}{2} = \sqrt{\frac{1 + \cos\alpha}{2}}.$$

[1] That is, the length of the perpendicular from the center to one of its sides.

EXAMPLE 6. Find rules for computing the radii of the inscribed and circumscribed circles of a regular 2^n-gon when its perimeter P is given.

SOLUTION. 1. $r_2 = \dfrac{P}{8}$, $R_2 = \dfrac{P\sqrt{2}}{8}$.

2. Knowing the radii r_n and R_n of the inscribed and circumscribed circles of a regular 2^n-gon of perimeter P, let us compute the radii r_{n+1} and R_{n+1} of the inscribed and circumscribed circles of a 2^{n+1}-gon having the same perimeter. Let AB (Fig. 1) be the side of a regular 2^n-gon of perimeter P, O be its center, C be the mid-point of the arc AB, and D be the mid-point of the chord AB. Further, let EF be the line joining the mid-points of the line segments AC and CB respectively, and let G be the mid-point of the line segment EF.

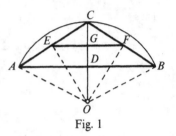

Fig. 1

Since

$$\angle EOF = \angle EOC + \angle FOC$$
$$= \frac{1}{2} \angle AOC + \frac{1}{2} \angle BOC$$
$$= \frac{1}{2} \angle AOB,$$

EF is equal to the side of a regular 2^{n+1}-gon inscribed in a circle of radius OE. The perimeter of this 2^{n+1}-gon is

$$2^{n+1}EF = 2^{n+1}\frac{AB}{2} = 2^n AB;$$

that is, it is also equal to P.

Thus,

$$r_{n+1} = OG \text{ and } R_{n+1} = OE.$$

But

$$r_n = OD \text{ and } R_n = OC.$$

Further, it is clear that $OC - OG = OG - OD$; hence,

$$R_n - r_{n+1} = r_{n+1} - r_n.$$

From this it follows that

$$r_{n+1} = \frac{R_n + r_n}{2}.$$

From the right triangle OEC, we have

$$OE^2 = OC \cdot OG,$$

or

$$R_{n+1}{}^2 = R_n r_{n+1}.$$

Hence,

$$R_{n+1} = \sqrt{R_n r_{n+1}}.$$

Thus, the required rules are

$$r_{n+1} = \frac{R_n + r_n}{2} \text{ and } R_{n+1} = \sqrt{R_n r_{n+1}}.$$

Now let us examine the sequence

$$r_2, R_2, r_3, R_3, \ldots, r_n, R_n, \ldots.$$

The terms of this sequence tend to the radius of a circle of perimeter P, that is, to $\frac{P}{2\pi}$. Specifically, for $P = 2$, this limit is $\frac{1}{\pi}$. We have

$$r_2 = \frac{1}{4} \text{ and } R_2 = \frac{\sqrt{2}}{4}.$$

Putting $r_1 = 0$ and $R_1 = \frac{1}{2}$, we obtain the following theorem:

If a sequence is composed of the terms

$$0, \quad \frac{1}{2}, \quad \frac{1}{4}, \quad \frac{\sqrt{2}}{4}, \quad \frac{\sqrt{2}+1}{8}, \quad \frac{\sqrt{2\sqrt{2}+4}}{8},$$

$$\frac{\sqrt{2\sqrt{2}+4}+\sqrt{2}+1}{16}, \ldots,$$

the first two terms of which are 0 and $\frac{1}{2}$, and each of the remaining terms of which is alternately equal to the arithmetic and geometric means of the two preceding terms, then the terms of this sequence approach $\frac{1}{\pi}$.

5. EXAMPLES PERTAINING TO *n*-GONS AND THEIR DIAGONALS

EXAMPLE 7. Find the sum of the interior angles of an *n*-gon (not necessarily convex!).

SOLUTION. 1. The sum of the interior angles of a triangle is $2 \cdot 90°$. The sum of the interior angles of a quadrilateral is $4 \cdot 90°$, since every quadrilateral may be divided into two triangles (Fig. 2).

2. Let us assume that we have proved that the sum of the interior angles of an arbitrary *k*-gon is equal to $2(k - 2) \cdot 90°$ for $k < n$, and let us consider the *n*-gon $A_1 A_2 \ldots A_n$.

First let us prove that for every polygon having more than three sides, it is possible to find a diagonal[1] which divides it into two polygons, each with a smaller number of sides (for a convex polygon this is obvious). Let A, B, C be

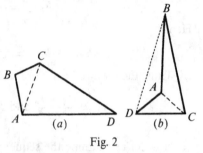

Fig. 2

any three adjacent vertices of the polygon. Then through the vertex *B* let us draw all possible half-lines filling the interior angle *ABC* of the polygon, and extend each half-line until it intersects the boundary of the polygon. Two cases may arise:

Case 1. All the half-lines intersect the same side of the polygon (Fig. 3*a*). In this case the diagonal *AC* divides our *n*-gon into an $(n - 1)$-gon and a triangle.

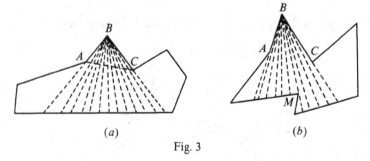

Fig. 3

Case 2. Not all of the half-lines intersect the same side of the polygon (Fig. 3*b*). In this case one of the half-lines will go through

---

[1] Notice that a diagonal of a nonconvex polygon may intersect it, or it may lie completely outside it (as illustrated by the diagonal *BD* in Fig. 2*b*).

some vertex M of the polygon, and the diagonal BM will divide the polygon into two polygons, each with a smaller number of sides.

Let us return now to the proof of our basic assertion. In the n-gon $A_1A_2 \ldots A_n$, let us draw a diagonal A_1A_k dividing it into a k-gon $A_1A_2 \ldots A_k$ and an $(n - k + 2)$-gon $A_1A_kA_{k+1} \ldots A_n$. According to the induction hypothesis, the sum of the interior angles of the k-gon is $(k - 2) \cdot 180°$, and that of the $(n - k + 2)$-gon is $[(n - k + 2) - 2] \cdot 180° = (n - k) \cdot 180°$; therefore the sum of the interior angles of an n-gon $A_1A_2 \ldots A_n$ will be equal to

$$(k - 2) \cdot 180° + (n - k) \cdot 180° = (n - 2) \cdot 180°,$$

from which it follows that our assertion is valid for all n.

As we have seen in Example 7, for any polygon it is possible to find a diagonal which divides it into two polygons, each with a smaller number of sides. Each of these polygons not already a triangle may again be divided into two polygons with fewer sides, and so on. Consequently, every polygon may be divided by nonintersecting diagonals into triangles.

EXAMPLE 8. Into how many triangles can an n-gon (not necessarily convex) be divided by its nonintersecting diagonals?

SOLUTION. 1. For a triangle this number is one (no diagonals can be drawn); for a quadrilateral this number is clearly two (see Fig. 2a, b).

2. Let us assume that we already know that every k-gon, where $k < n$, can be divided into $k - 2$ triangles by nonintersecting diagonals (irrespective of the method of division). Let us examine any one of the divisions of the n-gon $A_1A_2 \ldots A_n$ into triangles. Let A_1A_k be one of the diagonals of this division; it divides the n-gon $A_1A_2 \ldots A_n$ into a k-gon $A_1A_2 \ldots A_k$ and an $(n - k + 2)$-gon $A_1A_kA_{k+1} \ldots A_n$. On the basis of our assumption for $k < n$, the total number of triangles resulting from the division will be

$$(k - 2) + [(n - k + 2) - 2] = n - 2,$$

and thus our assertion is proved for all n.

Problem 4. Determine the number N of nonintersecting diagonals needed to divide an n-gon into triangles.

Hint. Since N diagonals and n sides of the n-gon appear as sides of $n - 2$ triangles (see Example 8), it follows that $2N + n = 3(n - 2)$, $N = n - 3$.

EXAMPLE 9. Show the method of computing $P(n)$, the number of ways in which a convex n-gon can be divided into triangles by non-intersecting diagonals.

SOLUTION. 1. For a triangle this is clearly one: $P(3) = 1$.

2. Let us assume that we have already determined the number $P(k)$ for all $k < n$; let us find $P(n)$ in this case. To do this, let us examine the convex n-gon $A_1A_2 \ldots A_n$ (Fig. 4). For any division of

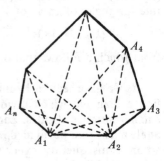

it into triangles the side A_1A_2 will be a side of one of the triangles of the division, and the third vertex of this triangle may be any one of the remaining points $A_3, A_4, \ldots, A_n$. The number of ways of dividing the n-gon into triangles so that this vertex coincides with the point A_3 is equal to the number of ways of dividing the $(n - 1)$-gon $A_1A_3A_4 \ldots A_n$ into triangles, that is, $P(n - 1)$. The number of ways of dividing the n-gon for which this vertex coin-

Fig. 4

cides with A_4 is equal to the number of ways of dividing the $(n - 2)$-gon $A_1A_4A_5 \ldots A_n$, that is, $P(n - 2) = P(n - 2)P(3)$. The number of ways of dividing for which this vertex coincides with A_5 is $P(n - 3)P(4)$, since each of the divisions of the $(n - 3)$-gon $A_1A_5 \ldots A_n$ may be combined with each of the divisions of the quadrilateral $A_2A_3A_4A_5$, and so on. Continuing in this way, we arrive at the following relationship:

$$P(n) = P(n - 1) + P(n - 2)P(3) + P(n - 3)P(4) + \cdots$$
$$+ P(3)P(n - 2) + P(n - 1). \tag{7}$$

Using this formula successively, we obtain

$P(4) = P(3) + P(3) = 2,$

$P(5) = P(4) + P(3)P(3) + P(4) = 5,$

$P(6) = P(5) + P(4)P(3) + P(3)P(4) + P(5) = 14,$

$P(7) = P(6) + P(5)P(3) + P(4)P(4) + P(3)P(5) + P(6) = 42,$

$P(8) = P(7) + P(6)P(3) + P(5)P(4) + P(4)P(5) + P(3)P(6) + P(7)$
$= 132,$ etc.

Remark.[1] Using formula (7), it is possible to prove that for any n

$$P(n) = \frac{2(2n - 5)!}{(n - 1)!(n - 3)!}.$$

Problem 5. Into how many pieces is a convex n-gon divided by all of its diagonals, if no three of them intersect at a common point?

Hint. The diagonal A_1A_n divides the convex $(n + 1)$-gon $A_1A_2 \ldots A_nA_{n+1}$ into an n-gon $A_1A_2 \ldots A_n$ and a triangle $A_1A_nA_{n+1}$. If we consider as known the number $F(n)$ of parts into which the n-gon $A_1A_2 \ldots A_n$ is divided by its diagonals, we can compute the number of additional pieces which result from the addition of the vertex A_{n+1}. This number is one more than the number of pieces into which the diagonals issuing from the vertex A_{n+1} are divided by the remaining diagonals. From vertex A_{n+1} there are $n + 1 - 3$, that is, $n - 2$, diagonals, and the number of pieces of each diagonal is one more than the number of intersecting diagonals.

Diagonal $A_{n+1}A_2$ is cut into $1 + (n - 2)$ parts;
diagonal $A_{n+1}A_3$ is cut into $1 + 2(n - 2) - 2$, or $1 + 2(n - 3)$ parts;
diagonal $A_{n+1}A_4$ is cut into $1 + 3(n - 2) - 3(2)$, or $1 + 3(n - 4)$ parts;
and so on.

Hence,

$$F(n + 1) = F(n) + 1 + [1 + (n - 2)] + [1 + 2(n - 3)] + \cdots \\ + \underbrace{[1 + (n - 3)2] + [1 + (n - 2)1]}_{n - 2 \text{ diagonals}}$$

$$= F(n) + 1 + (n - 2) + 1(n - 2) + 2(n - 3) + \cdots \\ + \underbrace{(n - 3)2 + (n - 2)1.}_{n - 2 \text{ terms}}$$

By using formulas (2) and (5) of the Introduction, we find

$$F(n + 1) = F(n) + (n - 1) + \frac{n(n - 1)(n - 2)}{6}$$

$$= F(n) + \frac{n^3}{6} - \frac{n^2}{2} + \frac{4n}{3} - 1.$$

Adding the values of $F(n)$, $F(n - 1)$, $\ldots$, $F(4)$, and using formulas (2), (3), and (4) of the Introduction, we obtain

$$F(n) = \frac{(n - 1)(n - 2)(n^2 - 3n + 12)}{24}.$$

---

[1] See the solution to Problem 51(*b*) in A. M. Yaglom and I. M. Yaglom, *Nonelementary Problems in Elementary Exposition*, Vol. I (San Francisco: Holden-Day, Inc., in press).

2. Proof by Induction

Certain of the propositions given in the preceding section can be regarded as examples of the use of the method of mathematical induction in proving geometric theorems. For example, the proposition of Example 7 could be formulated as follows: Prove that the sum of the angles of an n-gon is equal to $(n - 2) \cdot 180°$. In Example 8 it was proved that the nonintersecting diagonals divide an n-gon into $(n - 2)$ triangles. In this section we shall examine additional examples of the same type.

6. DISSECTIONS

EXAMPLE 10. Given n arbitrary squares, prove that it is possible to cut them into pieces such that these pieces can be rearranged to form a new square.

Proof. 1. For $n = 1$, our assertion does not require a proof. Let us prove that the assertion is valid for $n = 2$ as well. Let us denote the sides of the given squares $ABCD$ and $abcd$ by x and y, respectively; let $x \geq y$. On the sides of the square $ABCD$ with side x (Fig. 5a) let us lay off segments $AM = BN = CP = DQ = (x + y)/2$

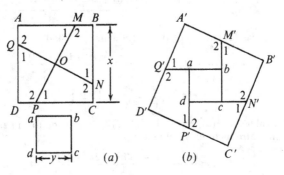

Fig. 5

and cut this square along the straight lines MP and NQ, which, as it is easy to see, intersect each other at the center O of the square and form a right angle, dividing the square into four congruent

16

pieces. Let us arrange these pieces around the second square as shown in Fig. 5b. The figure obtained is also a square, as the angles at the points M', N', P', and Q' are supplementary, the angles at A', B', C', and D' are right angles, and $A'B' = B'C' = C'D' = D'A'$.

2. Let us assume that our assertion has been proved for n squares, and let $n + 1$ squares $K_1, K_2, \ldots, K_n, K_{n+1}$ be given. Let us choose any two of these squares, say K_n and K_{n+1}. As shown in paragraph 1, we can cut one of these squares into pieces and adjoin these pieces to the other square to form a new square K'. Further, according to our assumption, the squares $K_1, K_2, \ldots, K_{n-1}, K'$ can be cut into pieces in such a way that these pieces may be arranged to form a new square.

EXAMPLE 11. Given a triangle ABC. Through its vertex C there are drawn $n - 1$ straight lines $CM_1, CM_2, \ldots, CM_{n-1}$ dividing the triangle into n smaller triangles $ACM_1, M_1CM_2, \ldots, M_{n-1}CB$. Let $r_1, r_2, \ldots, r_n$ and $\rho_1, \rho_2, \ldots, \rho_n$ denote, respectively, the radii of the inscribed circles of these triangles and the escribed circles that fall within the angle C of each triangle (see Fig. 6a). Let r and ρ be the

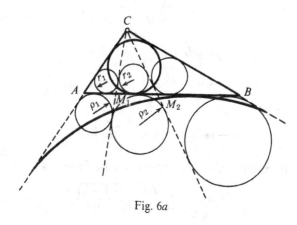

Fig. 6a

radii of the inscribed and escribed circles of the triangle ABC itself. Prove that

$$\frac{r_1}{\rho_1} \cdot \frac{r_2}{\rho_2} \cdot \ldots \cdot \frac{r_n}{\rho_n} = \frac{r}{\rho}.$$

Proof. Let $S(ABC)$ denote the area of the triangle ABC and let s denote half its perimeter; then, as is well known, $S(ABC) = sr$.

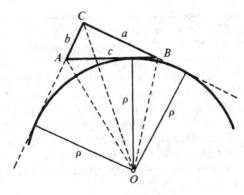

Fig. 6*b*

On the other hand, if O is the center of the escribed circle of this triangle (Fig. 6*b*), then

$$S(ABC) = S(OAC) + S(OCB) - S(OAB)$$

$$= \frac{1}{2}b\rho + \frac{1}{2}a\rho - \frac{1}{2}c\rho$$

$$= \frac{1}{2}(b + a - c)\rho$$

$$= (s - c)\rho;$$

consequently,

$$sr = (s - c)\rho \quad \text{and} \quad \frac{r}{\rho} = \frac{s - c}{s}.$$

Further, from the well-known trigonometric formulas

$$\tan \frac{A}{2} = \sqrt{\frac{(s - b)(s - c)}{s(s - a)}}$$

and

$$\tan \frac{B}{2} = \sqrt{\frac{(s - a)(s - c)}{s(s - b)}},$$

we have

$$\tan \frac{A}{2} \tan \frac{B}{2} = \sqrt{\frac{(s - b)(s - c)}{s(s - a)} \times \frac{(s - a)(s - c)}{s(s - b)}} = \frac{s - c}{s}.$$

But

$$\frac{s-c}{s} = \frac{r}{\rho},$$

and so

$$\tan \frac{A}{2} \tan \frac{B}{2} = \frac{r}{\rho}. \tag{8}$$

After these preliminary remarks let us turn to the proof of the theorem.

1. For $n = 1$ there is nothing to prove. Let us show that our assertion is valid for $n = 2$. In this case the line CM divides the triangle ABC into two smaller triangles ACM and CMB. By formula (8)

$$\begin{aligned}
\frac{r_1}{\rho_1} \cdot \frac{r_2}{\rho_2} &= \tan \frac{A}{2} \tan \frac{CMA}{2} \tan \frac{CMB}{2} \tan \frac{B}{2} \\
&= \tan \frac{A}{2} \tan \frac{CMA}{2} \tan \frac{180° - CMA}{2} \tan \frac{B}{2} \\
&= \tan \frac{A}{2} \tan \frac{B}{2} \\
&= \frac{r}{\rho}.
\end{aligned}$$

2. Let us assume that our assertion has been proved for $n - 1$ straight lines, and let n straight lines $CM_1, CM_2, \ldots, CM_n$ be given, dividing the triangle ABC into $n + 1$ smaller triangles $ACM_1, M_1CM_2, \ldots, M_nCB$. Let us consider two of these triangles, say ACM_1 and M_1CM_2. As we have seen in paragraph 1,

$$\frac{r_1}{\rho_1} \cdot \frac{r_2}{\rho_2} = \frac{r_{12}}{\rho_{12}},$$

where r_{12} and ρ_{12} are the radii of the inscribed and escribed circles of the triangle ACM_2. But by virtue of our assumption, for the n triangles $ACM_2, M_2CM_3, \ldots, M_nCB$ the equality

$$\frac{r_{12}}{\rho_{12}} \cdot \frac{r_3}{\rho_3} \cdot \ldots \cdot \frac{r_n}{\rho_n} \cdot \frac{r_{n+1}}{\rho_{n+1}} = \frac{r}{\rho}$$

holds, and consequently,

$$\frac{r_1}{\rho_1} \cdot \frac{r_2}{\rho_2} \cdot \ldots \cdot \frac{r_n}{\rho_n} \cdot \frac{r_{n+1}}{\rho_{n+1}} = \frac{r}{\rho}.$$

Problem 6. Let the straight lines CM and CM' divide the triangle ABC in two different ways into two triangles, ACM, CMB and ACM', $CM'B$; let r_1, r_2 and r_1', r_2' be, respectively, the radii of the circles inscribed in these triangles. Prove that if $r_1 = r_1'$, then $r_2 = r_2'$, and that an analogous statement holds for the radii of the escribed circles also.

Hint. Using the notation of Example 11, prove first that

$$\frac{r}{\rho} = 1 - \frac{2r}{h} \quad \text{and} \quad \frac{\rho}{r} = 1 + \frac{2\rho}{h}$$

(where h is the altitude from the vertex C), from which follow the equalities

$$\left(1 - \frac{2r_1}{h}\right)\left(1 - \frac{2r_2}{h}\right) = 1 - \frac{2r}{h} = \left(1 - \frac{2r_1'}{h}\right)\left(1 - \frac{2r_2'}{h}\right)$$

and

$$\left(1 + \frac{2\rho_1}{h}\right)\left(1 + \frac{2\rho_2}{h}\right) = 1 + \frac{2\rho}{h} = \left(1 + \frac{2\rho_1'}{h}\right)\left(1 + \frac{2\rho_2'}{h}\right).$$

Problem 7. In the notation of Example 11 prove that

$$\frac{r_1 + \rho_1}{R_1} + \frac{r_2 + \rho_2}{R_2} + \cdots + \frac{r_n + \rho_n}{R_n} = \frac{r + \rho}{R},$$

where R_1, R_2, ..., R_n, and R are the radii of the circumscribed circles of the triangles ACM_1, M_1CM_2, ..., M_nCB, and ABC, respectively.

Hint. As is well known,

$$S(ABC) = sr = (s - c)\rho = \frac{abc}{4R}, \quad \text{and}$$

$$S(ABC) = \sqrt{s(s - a)(s - b)(s - c)},$$

whence we obtain

$$\frac{r + \rho}{2R} = \frac{\dfrac{S(ABC)}{s} + \dfrac{S(ABC)}{s - c}}{\dfrac{abc}{2S(ABC)}}$$

$$= \frac{(a + b)\,[c^2 - (a - b)^2]}{2abc}$$

$$= \frac{b^2 + c^2 - a^2}{2bc} + \frac{a^2 + c^2 - b^2}{2ac}$$

$$= \cos CAB + \cos CBA,$$

using the law of cosines.

Problem 8. Given n circles C_1, $C_2, \ldots, C_n$ passing through the point O, let us denote the second points of intersection of the circles C_1 and C_2, C_2 and $C_3, \ldots, C_n$ and C_1 by A_1, $A_2, \ldots, A_n$, respectively (Fig. 7a). Let B_1 be an arbitrary point on the circle C_1,

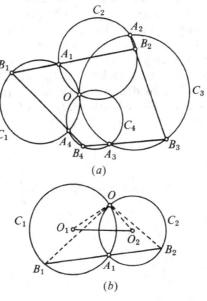

(a)

(b)

Fig. 7

different from O and from A_1. Let us draw the secant B_1A_1, intersecting the circle C_2 at the point B_2; then the secant B_2A_2, intersecting the circle C_3 at the point B_3; and so on. (In case the point B_2 coincides with A_2, for example, instead of the secant through the point A_2, we draw the tangent to the circle C_2.) Prove that the point B_{n+1} finally obtained on the circle C_1 will coincide with B_1.

Hint. First prove the following lemma: let O_1 and O_2 be the centers of the circles C_1 and C_2 intersecting at the point O, and let B_1B_2 be a secant drawn through the point A_1, the second point of intersection of these two circles (see Fig. 7b); then the segments B_1B_2 and O_1O_2 as viewed from the point O subtend the same angle. Next prove the proposed theorem for three circles. After this is done, assuming that the theorem is valid for $n-1$ circles, consider the case for n circles C_1, $C_2, \ldots, C_n$. Draw the secant through the point B_{n-1} and the point of intersection of the circles C_{n-1} and C_1; apply the inductive hypothesis to the $n-1$ circles C_1, $C_2, \ldots, C_{n-1}$.

7. MAPS

Suppose that we are given a network of lines in the plane which connect certain points $A_1, A_2, \ldots, A_p$ and have no other points in common. We shall assume further that this network of lines "consists of a single piece," that is, that it is possible to start from any one of the points $A_1, A_2, \ldots, A_p$ and to get to any other point, moving only along the lines of the network. (This is the property of *connectedness*.) Such a network of lines will be called a *map*, the given points its *vertices*, the segments of the curves between pairs of adjoining vertices the *boundaries* of the map, the pieces of the plane into which it is divided by the boundaries (including the infinite exterior domain as well) the *countries* of the map. Thus, in Fig. 8, the points $A_1, A_2, A_3, A_4, A_5, A_6, A_7, A_8$ are the vertices of the

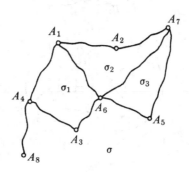

Fig. 8

map; the lines $A_1A_2, A_2A_7, A_1A_6, A_6A_7, A_4A_1, A_4A_3, A_3A_6, A_6A_5, A_5A_7, A_4A_8$ are its boundaries; and the regions $\sigma_1, \sigma_2, \sigma_3$, and the infinite exterior region σ are its countries.

EXAMPLE 12. (EULER'S THEOREM) In an arbitrary map[1] let s denote the number of countries, l the number of boundaries, and p the number of vertices. Then

$$s + p = l + 2.$$

Proof. Let us prove this theorem by induction on l, the number of boundaries in the map.

[1] All maps are assumed to have at least one vertex.

1. Let $l = 0$; then $s = 1$ and $p = 1$; in this case

$$s + p = l + 2.$$

2. Let us assume that the theorem is valid for an arbitrary map having n boundaries and consider a map containing $l = n + 1$ boundaries, s countries, and p vertices. Two cases are possible.

Case 1. Between any two vertices of the map there is *only one* path along the boundaries of the map connecting the two points. (There is always at least one such path in view of the connectedness of the map.) In this case the map does not contain a single closed contour, and, consequently, has the form illustrated in Fig. 9; for this map $s = 1$. We shall show that for such a map, one can

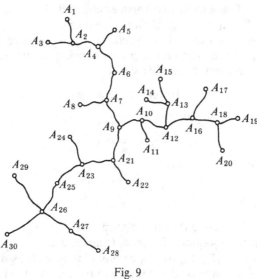

Fig. 9

find at least one vertex which is the end of only one boundary. (Such is the vertex A_1 in Fig. 9; such a vertex will be called an *extremity*.) Indeed, let us select an arbitrary vertex of the map. If it is not an extremity, then at least two boundaries end at this vertex. Let us proceed along one of these boundaries to a second vertex. If this vertex is not an extremity either, then it is the end point of some other boundary; let us proceed along this boundary to its other end point, and so on. Since by hypothesis the map does not contain any closed contours, we shall not return to any of the earlier vertices, and because there are only a finite number of vertices in the map, we must eventually come to a vertex which will be an extremity. Removing this vertex together with the single boundary which ends in this vertex, we obtain a new map for which

$$l' = l - 1 = n, \quad s' = s = 1, \quad p' = p - 1,$$

this new map, of course, remaining connected. By virtue of the inductive hypothesis we have

$$s' + p' = l' + 2,$$

from which we have

$$s + p = l + 2.$$

Case 2. There exist two vertices connected by more than one path (Fig. 8). In this case the map contains some closed contour which passes through these vertices. If we remove one of the boundaries of this contour (without removing the vertices), we obtain a new connected map for which

$$l' = l - 1 = n, \quad p' = p, \quad s' = s - 1.$$

By the inductive hypothesis

$$s' + p' = l' + 2,$$

from which we have

$$s + p = l + 2.$$

EXAMPLE 13. Prove that if the map contains no vertex at which fewer than three boundaries meet (that is, if the map contains no vertices such as A_2, A_3, A_5, A_8 or boundaries such as A_4A_8 in Fig. 8), then there is at least one country in the map which has no more than five boundaries.

Proof. Since at least three boundaries meet in each of the p vertices of the map, then $3p$ will not exceed twice the number of boundaries $2l$ (the number is doubled, since each boundary connects two vertices); hence

$$p \leq \frac{2}{3} l. \tag{9}$$

Let us assume now that each of the s countries of the map has no less than six boundaries; then $6s$ does not exceed twice the number of boundaries $2l$ (doubled because each boundary separates two countries), from which we have

$$s \leq \frac{1}{3} l. \tag{10}$$

Inequalities (9) and (10) give

$$s + p \leq \frac{1}{3} l + \frac{2}{3} l = l,$$

which contradicts Euler's theorem. Consequently, our supposition that each country has no less than six boundaries is incorrect.

Problem 9. Let five points be given in the plane. Prove that it is impossible to join every pair of these points with noninter-secting lines (Fig. 10).

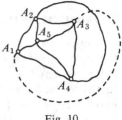

Fig. 10

Hint. Assuming that all these points are joined according to the conditions of the prob-lem, we arrive at a map with 5 vertices, $\frac{5 \cdot 4}{2} = 10$ boundaries, and, consequently, 7 countries (Euler's theorem!). The impossibility of such a map follows from considerations similar to those which led to inequality (10).[1]

8. POLYHEDRONS

Problem 10. (EULER'S THEOREM FOR POLYHEDRONS) Prove that if p is the number of vertices, l the number of edges, and s the num-ber of faces in a convex polyhedron, then

$$s + p = l + 2.$$

Hint. Place the polyhedron inside a sphere of sufficiently large radius, and from the center of the sphere (which we may assume lies inside the poly-hedron) project all points of the polyhedron onto the surface of the sphere. This gives a map on the surface of the sphere. From an arbitrary point on the surface of the sphere which is not a point on any boundary of the map, project the map onto a plane tangent to the sphere at a point diametrically opposed to the first one (*stereographic projection*). Apply Euler's theorem to the plane map thus obtained.

---

[1] The reader can find other examples of the application of Euler's theorem on maps in a book by E. B. Dynkin and V. A. Uspenskii, *Multicolor Problems,* translated by N. D. Whaland, Jr., and R. B. Brown (Boston: D. C. Heath and Company, 1963).

Problem 11. Prove that every polyhedron has at least one face that is either a triangle, a quadrilateral, or a pentagon.

Hint. See Example 13.

Problem 12. Prove that there does not exist a polyhedron with seven edges.

Hint. Apply Euler's theorem.

Remark. Other examples of the applications of Euler's theorem for polyhedrons may be found, for example, in the book by R. Courant and H. Robbins, *What Is Mathematics?* (New York: Oxford University Press, 1941).

9. PROBLEMS OF MAP COLORING

Suppose we are given some map in the plane. We shall say that the map is *colored properly* if each of its countries has a definite color and any two countries having a common boundary between them are colored with different colors. Examples of properly colored maps may be found by looking at any geographical map. Any map may be properly colored, for example, by coloring each country a special color, but such coloring is not economical. The question naturally arises: What is the *smallest* number of colors necessary to color a given map properly? It is clear that, for example, the map shown in Fig. 11*a* may be colored properly with two colors; to color properly the map shown in Fig. 11*b*, we need three colors; and the map shown in Fig. 11*c* may be colored properly only by using four

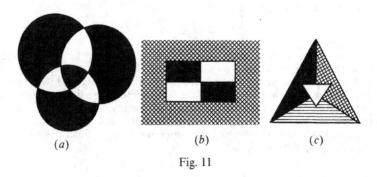

(*a*)   (*b*)   (*c*)

Fig. 11

colors. To date, no one has ever found a single map which could not be properly colored with four colors. Apparently the famous

German mathematician Möbius was the first to notice this circumstance, more than a hundred years ago. Since that time many great scholars have attempted to solve this *four-color problem,* that is, either to prove that four colors are sufficient to color any map properly or to find an example of a map for which four colors are insufficient, but to this day no one has succeeded in doing either. It has been established, however, that *five* colors are sufficient to color any map properly (see Example 17 below). Nor is it difficult to find conditions under which a map can be colored with two colors (Example 15) or three colors (Example 16). We shall also point out a certain condition which is necessary and sufficient for a map to be properly colored with four colors (Example 18); but the question whether this condition is satisfied for an arbitrary map, or whether there are maps which do not satisfy this condition, remains open, of course.

It is interesting to note that for certain surfaces which would appear to be more complicated than the plane, the map coloring problem has been completely solved. Thus, for example, it has been proved that on a "doughnut," or torus (Fig. 12), seven colors are sufficient to color any map properly, while for some maps six colors are insufficient for proper coloring.

Fig. 12

In what follows we shall assume that the maps contain no boundaries which do not serve as separations, that is, boundaries such that one and the same country lies on both sides (such as the boundary A_4A_8 in Fig. 8); otherwise the statement of the problem of proper coloring does not make sense. We shall assume also that the map does not contain vertices at which only two boundaries meet (such as the vertex A_2 in Fig. 8), because such vertices would be superfluous. In other words, we shall consider only those maps for which at each vertex at least three boundaries meet, that is, maps which satisfy the condition of Example 13, and in what follows the result obtained in that example will be used repeatedly. Also, it will be convenient for us to assume that the maps have only one infinite domain, that is, that the maps have no boundaries which "recede to infinity." It can be shown that to give up this last condition would not alter any of the deductions which follow.

10. NORMAL MAPS

We shall call a map *normal* if exactly three boundaries meet at each vertex. Suppose that we are given an arbitrary map S (as in Fig. 13a). Let us draw a small circle around each vertex of this map at

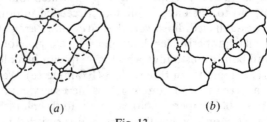

(a)   (b)

Fig. 13

which more than three boundaries meet, and then add the region bounded by this circle to one of the countries which surround this vertex. In this way we obtain a normal map S' with the same number of countries (as in Fig. 13b); from any proper coloring of the map S' it is easy to obtain a proper coloring of the map S using the same number of colors.[1] Therefore the four-color problem for all maps is equivalent to the four-color problem for normal maps. Hence, we shall often restrict our attention to normal maps.

We shall now look into the construction of some of the simplest normal maps.[2] Let p be the number of vertices, l the number of boundaries, and s the number of countries in a normal map; then $2l = 3p$ (see above, Example 13), and $p = \frac{2}{3} l$. Moreover, by Euler's theorem $s + p = l + 2$; hence,

$$s = (l - p) + 2 = \frac{l}{3} + 2,$$

and, consequently, $s \geq 2$. But for $s = 2$, we find that $l = 0$; obviously no such map exists. Putting $s = 3$, we obtain $l = 3$ and $p = 2$; this simple normal map has the form illustrated in Fig. 14a. For $s = 4$, we obtain $l = 6$ and $p = 4$. We shall show that in this case the map has the form illustrated in Fig. 14b or 14c. Indeed,

---

[1] The converse assertion is false. It is not true that a proper coloring of S always leads to a proper coloring of S'.

[2] Here and in what follows we do not distinguish between "similarly arranged" maps (as illustrated in Fig. 11c and 23a), maps in which the countries and boundaries can be numbered so that in both maps countries bearing the same number are separated by boundaries bearing the same number.

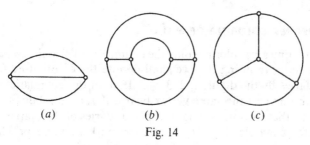

(a) (b) (c)

Fig. 14

let k_2 denote the number of countries with two vertices, k_3 the number of countries with three vertices, and k_4 the number of countries with four vertices (since $p = 4$, the map cannot have countries with more than four vertices). Then

and
$$k_2 + k_3 + k_4 = s = 4$$
$$2k_2 + 3k_3 + 4k_4 = 2l = 12$$

(see above, Example 13). From the latter equality it is clear that k_3 must be even. The sum of $k_2 + k_3 + k_4$, equal to 4, with the terms arranged in decreasing order may have the form $2 + 2 + 0$, $2 + 1 + 1$, $3 + 1 + 0$, or $4 + 0 + 0$. Let us examine each of these cases.

If two of the values of k are equal to 2 and one to 0, then for $k_2 = 2$, $k_3 = 2$, $k_4 = 0$ the sum $2k_2 + 3k_3 + 4k_4 = 10$, which is impossible because $2k_2 + 3k_3 + 4k_4 = 12$; for $k_2 = 2$, $k_3 = 0$, $k_4 = 2$ the sum $2k_2 + 3k_3 + 4k_4 = 12$, and this case corresponds to the map represented in Fig. 14b; for $k_2 = 0$, $k_3 = 2$, $k_4 = 2$ the sum $2k_2 + 3k_3 + 4k_4 = 14 > 12$.

If one of the values of k is 2 and the other two are 1, then, since k_3 must be even, we can have only $k_2 = 1$, $k_3 = 2$, $k_4 = 1$, for which the sum $2k_2 + 3k_3 + 4k_4 = 12$. Such a map exists, but it is not a normal map (Fig. 15).

If one of the values of k is 3 and another is 1, then k_3 must be equal to 0, since k_3 is even; in this case $2k_2 + 4k_4 \neq 12$.

Finally, if one of the values of k is 4 and the remaining values are 0, then only for $k_2 = k_4 = 0$, $k_3 = 4$ is the sum $2k_2 + 3k_3 + 4k_4 = 12$. The corresponding map has the form shown in Fig. 14c.

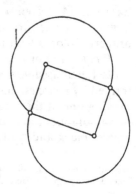

Fig. 15

11. PROPER COLORING OF MAPS

Sometimes we color not only the countries but also the *boundaries* of the map; in doing this we shall denote the colors used for the boundaries by the digits, 1, 2, 3, ... If all of the boundaries which meet in one and the same vertex bear different numbers, then we shall say that the numbering of the boundaries of the map is *proper* (see, for example, Fig. 16). Let us remark that the problem of

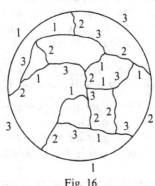

Fig. 16

numbering the vertices of a map so that "neighboring" vertices, that is, vertices connected by the same boundary, have different numbers, is connected with the problem of proper coloring of the countries of a map.[1]

EXAMPLE 14. Given n circles in the plane, prove that no matter how these circles are arranged, the map which they form may be properly colored with two colors.

Proof. 1. For $n = 1$, the assertion is obvious.

2. Let us assume that our assertion is valid for an arbitrary map formed by n circles, and let $n + 1$ circles be given in the plane. Removing one of these circles, we obtain a map which, by virtue of our assumption, may be properly colored with two colors, for example, black and white (Fig. 17*a*). Now let us restore the deleted

[1] In this connection see, for example, the book by E. B. Dynkin and V. A. Uspenskii mentioned on p. 25. Here the reader will find other proofs of the theorems given below.

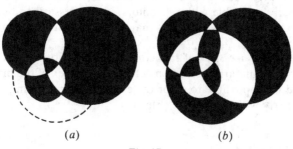

(a)   (b)

Fig. 17

circle, and on one side of it (for example, on the inside) let us change the color of each domain into its opposite (that is, black to white, and vice versa); it is easy to see that we thereby obtain a map that is properly colored with two colors (Fig. 17b).

Problem 13. We are given n circles in the plane, with a chord drawn in each. Prove that the map formed by these circles and chords can be properly colored with three colors (Fig. 18).

Fig. 18

Hint. Suppose that the map formed by n circles with chords can be properly colored with three colors α, β, γ.

Draw the $(n + 1)$st circle and change the colors of the countries lying inside this circle and on one side of the corresponding chord according to the scheme

$$\alpha \to \beta, \quad \beta \to \gamma, \quad \gamma \to \alpha,$$

and change the colors of the countries lying inside the circle and on the other side of the chord according to the scheme

$$\alpha \to \gamma, \quad \beta \to \alpha, \quad \gamma \to \beta.$$

EXAMPLE 15. (THE TWO-COLOR THEOREM) In order for a map to be properly colorable with two colors, it is necessary and sufficient that the number of boundaries that meet at each of its vertices be even.

Proof. The *necessity* of this condition is obvious, for if an odd number of boundaries were to meet at some vertex of the map, then it would be impossible to color the countries surrounding that vertex properly with two colors (Fig. 19).

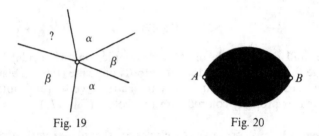

Fig. 19   Fig. 20

To prove the *sufficiency* of the condition let us carry out induction on the number of boundaries of the map.

1. For a map with two boundaries the assertion is obvious (Fig. 20).

2. Let us assume that the theorem is valid for an arbitrary map in which an even number of boundaries meet at each vertex, and for which the total number of boundaries does not exceed n, and let a map S be given having $n + 1$ boundaries and satisfying this condition. Beginning at an arbitrary vertex A of the map S, let us move in an arbitrary direction along a boundary of the map. Since there are a finite number of vertices on the map, we shall return finally to one of the preceding vertices. (The map has no extremities, since it has no boundaries which do not serve as separations.) We can thus single out a nonself-intersecting closed curve consisting of boundaries of the map. Deleting this curve, we obtain a map S' with fewer boundaries, but which also has an even number of boundaries meeting at each vertex (because at each vertex of the map S an even number of boundaries, either 0 or 2, has been removed). By the inductive hypothesis the map S' may be properly colored with two colors.

If we now restore the deleted curve and reverse every color on one side of the curve (for example, the inside), we shall have a proper coloring of the map S.

EXAMPLE 16. (THE THREE-COLOR THEOREM) In order for a normal map to be properly colorable with three colors, it is necessary and sufficient that each of its countries have an even number of boundaries.

Proof. The *necessity* of the formulated condition is obvious, because if the map has a country σ with an odd number of boundaries, then σ and the countries contiguous to it cannot be properly colored with three colors (Fig. 21).

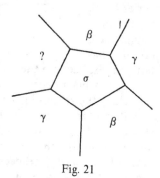

Fig. 21

To prove the *sufficiency* of the condition let us carry out induction on n, the number of countries in the map.

1. For a normal map with 3 countries (see Fig. 14a) our assertion is obvious. The normal map with 4 countries, shown in Fig. 14b, can obviously be colored properly with three colors also (to achieve this it is sufficient to use the same color for the "inner" country and the outer region). The normal map shown in Fig. 14c does not satisfy the condition that each country have an even number of boundaries. Thus, every normal map with 3 or 4 countries, each country of which has an even number of boundaries, may be properly colored with three colors.

2. Let us assume that the theorem is true for an arbitrary normal map each country of which has an even number of boundaries and of which the total number of countries is either $n - 1$ or n, and let us examine a normal map S which satisfies the same condition and has $n + 1$ countries. As we have shown in Example 13, such a map S will have a country σ with no more than 5 boundaries. Hence, σ will have either two or four boundaries. Let us examine each of these cases.

Case 1. σ has *two* boundaries. Let A and B be the vertices of this country and let σ_1 and σ_2 be the countries which adjoin (Fig. 22). Removing the boundary between the countries σ and σ_1, we obtain a map S' which is also normal, because the points A and B will no longer be vertices (we have agreed that a map has no superfluous vertices), and the number of boundaries which meet at each of the remaining vertices remains the same. Each country of the map S' also has

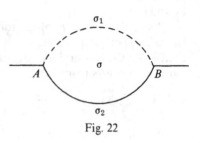

Fig. 22

an even number of boundaries, because the number of boundaries for the countries σ_1 and σ_2 has been decreased by 2, and the number of boundaries for each of the remaining countries remains unchanged. Since the number of countries in the map S' is n, by the inductive hypothesis it is possible to color S' properly with three colors α, β, and γ. Suppose the countries $\sigma_1' = \sigma_1 + \sigma$ and $\sigma_2' = \sigma_2$ receive the colors α and β, respectively. Restoring the country σ and coloring it with the color γ, we obtain a proper coloring for the map S.

Case 2. σ has *four* boundaries. Suppose that two of the countries adjacent to σ and lying on opposite sides of σ were to have a common boundary between them (Fig. 23a), or that two countries were to coincide (as σ_3 in Fig. 14b, in which σ_2 plays the role of σ in Fig. 23a). However, in these cases the two other countries adjacent to σ could neither have a common boundary nor coincide.

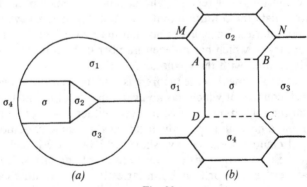

(a)   (b)

Fig. 23

Let countries σ_2 and σ_4 be these two other countries (Fig. 23*b*). Let us join the countries σ_2 and σ_4 to country σ, removing the boundaries AB and CD. In this way we obtain a map S' which is obviously also normal and on this map also each country will have an even number of boundaries. Indeed, if $2k_1$ is the number of boundaries in the country σ_1, $2k_2$ the number of boundaries in the country σ_2, $2k_3$ the number of boundaries in the country σ_3, and $2k_4$ the number of boundaries in the country σ_4, then the country $\sigma' = \sigma + \sigma_2 + \sigma_4$ will have $2k_2 + 2k_4 - 4$ boundaries, the country $\sigma_1' = \sigma_1$ will have $2k_1 - 2$ boundaries, and the country $\sigma_3' = \sigma_3$ will have $2k_3 - 2$ boundaries, while the number of boundaries of each of the remaining countries will remain unchanged. (In the case where countries σ_1 and σ_3 coincide, this country will have 4 boundaries less in the map S' than in the map S.)

Since map S' has $n - 1$ countries, then by the inductive hypothesis it is possible to color S' properly with three colors α, β, and γ. Let us show that in this case the countries σ_1' and σ_3' will be colored with the same color (this assertion is obvious if σ_3' coincides with σ_1'). Indeed, suppose country σ' is colored with the color α, and country σ_1' with the color β. Since along the portion MN of the boundary an odd number $(2k_2 - 3)$ of countries adjoin σ', and the coloring of these countries must obviously alternate in the sequence γ, β, γ, β, ..., γ, the country σ_3' must be colored β. Restoring the country σ and giving it the color γ, we obtain a proper coloring for the map S.

12. THE FIVE-COLOR THEOREM

EXAMPLE 17. (THE FIVE-COLOR THEOREM) Any normal map can be colored properly with five colors.

Proof. 1. If the number of countries in the map does not exceed five, the assertion is obvious.

2. Let us assume that the theorem is valid for an arbitrary normal map with $n - 1$ or n countries, and let us consider a map S consisting of $n + 1$ countries. As we have shown in Example 13, the map S contains at least one country σ for which the number of boundaries does not exceed 5. Let us consider all possible cases which can occur.

Case 1. σ has *two* boundaries (see Fig. 22). Let σ_1 and σ_2 be the countries adjacent to σ. Joining the country σ_1 to σ, we obtain a normal map S' with n countries.

By the inductive hypothesis, S' can be properly colored with five colors. The countries

$$\sigma_1' = \sigma + \sigma_1 \quad \text{and} \quad \sigma_2' = \sigma_2$$

will be colored with some two of these colors. Restoring country σ, we can color it with any one of the three remaining colors.

Case 2. σ has *three* boundaries (Fig. 24a). Let us join σ_1 to σ. If the resulting map S' is colored with five colors, we can then color the country σ with one of the two colors not used for coloring the countries

$$\sigma_1' = \sigma + \sigma_1, \quad \sigma_2' = \sigma_2, \quad \text{and} \quad \sigma_3' = \sigma_3.$$

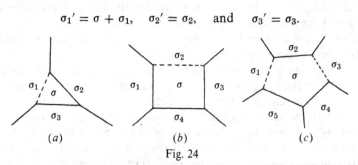

Fig. 24

Case 3. σ has *four* boundaries (Fig. 24b). It is possible to find two countries adjoining σ which do not coincide (see Example 16). Joining σ to one of these countries, for example σ_2, we obtain a map S' with n countries, which, by the inductive hypothesis, can be colored properly with five colors. The countries

$$\sigma_1' = \sigma_1, \quad \sigma_2' = \sigma_2 + \sigma, \quad \sigma_3' = \sigma_3, \quad \text{and} \quad \sigma_4' = \sigma_4$$

are colored with some four of the five possible colors (or fewer, if σ_1' and σ_3' coincide or have the same color). Restoring the country σ, we can color it with the fifth color.

Case 4. σ has *five* boundaries (Fig. 24c). As in Example 16, we can find two countries adjoining σ without a common boundary and not coinciding; let these countries be σ_1 and σ_3. Joining both of these countries to σ, we obtain a normal map S' with $n - 1$ countries. By the inductive hypothesis, the map S' can be properly colored with five colors. The countries

$$\sigma_1' = \sigma_1 + \sigma + \sigma_3, \quad \sigma_2' = \sigma_2, \quad \sigma_4' = \sigma_4, \quad \text{and} \quad \sigma_5' = \sigma_5$$

will be colored with four of the five colors. Restoring country σ, we can color it with the fifth color.

13. VOLYNSKII'S THEOREM

EXAMPLE 18. (VOLYNSKII'S THEOREM)[1] A normal map can be properly colored with four colors if and only if its boundaries can be properly numbered with three digits.[2]

Proof. **A.** If a normal map can be properly colored with four colors, then its boundaries can be properly numbered with three digits.

Suppose the normal map S is properly colored with four colors α, β, γ, and δ. Let us assign the digit 1 to boundaries between countries colored α and β, or γ and δ; the digit 2 to boundaries between countries colored α and γ, or β and δ; and the digit 3 to boundaries between countries colored α and δ, or β and γ. The numbering of the boundaries which results is normal. Indeed, if two boundaries bearing the same digit meet at any vertex A (for example, the digit 1, Fig. 25), then the countries σ_2 and σ_3, which are

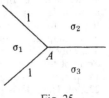

Fig. 25

separated from country σ_1 by the boundaries bearing the same digit, must have the same color (thus, if σ_1 in our example is colored α, then σ_2 and σ_3 will be colored β); but this cannot occur because σ_2 and σ_3 have a common boundary.

B. If the boundaries of a normal map can be properly numbered with three digits, then its countries can be properly colored with four colors. The proof of this will be given by induction on n, the number of countries.

1. The boundaries of the simplest normal map, having three countries (see Fig. 14a), can be numbered with the digits 1, 2, 3 in only one way (except for a permutation of these digits). Let us color this map as indicated in Fig. 26a. The boundary between the countries colored α and β will bear

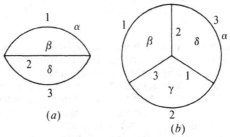

(a)

(b)

Fig. 26

---

[1] V. V. Volynskii (1923–1943), a Soviet mathematician.

[2] It appears that Volynskii's theorem was discovered earlier by P. G. Tait (Scientific Papers, Cambridge, 1898–1900, vol. 1, pp. 408–411 and vol. 2, pp. 85–98). Reference is made to this theorem of Tait's by König (*Graphentheorie*, Chelsea Publishing Co., New York, 1950, p. 202) and some details are given by W. W. R. Ball in *Mathematical Recreations and Essays*, revised by H. S. M. Coxeter (New York: The Macmillan Company, 1960). Tait's work gives the key idea of the proof, but with no details, and Ball's account, although more detailed, is still incomplete. It seems quite likely that Volynskii was unaware of Tait's work.

the number 1, the boundary between the countries colored β and δ the number 2, and the boundary between the countries colored α and δ the number 3.

We shall say that a proper coloring of a map S which is colored with four colors α, β, γ, δ is *admissible* if each boundary between colors α and β and between colors γ and δ bears the number 1, each boundary between colors α and γ and between colors β and δ bears the number 2, and each boundary between colors α and δ and between colors β and γ bears the number 3. We have shown that the simplest normal map with three countries can be given an admissible proper coloring with four colors. We shall show that this is also the case for a normal map with four countries (Fig. 14b, c). The boundaries of the map shown in Fig. 14c can be properly numbered with the digits 1, 2, 3 in only one way, except for a permutation of the digits (Fig. 26b). The coloring of this map, shown in Fig. 26b, is admissible. The map shown in Fig. 14b permits two essentially different ways of numbering the boundaries (Fig. 27a, b). The colorings of these maps, shown in Fig. 27a, b, are also admissible.

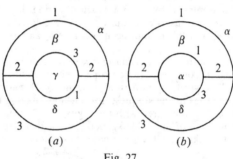

Fig. 27

2. Let us assume that for any normal map with $n - 1$ or n countries for which the boundaries are properly numbered with three digits, there is an admissible proper coloring with four colors, and let us examine a normal map S, having $n + 1$ countries for which the boundaries are also properly numbered with three digits. As we saw in Example 13, there is a country σ in S for which the number of boundaries does not exceed five. Let us consider the various particular cases which can occur.

Case 1. σ has *two* boundaries. Except for a permutation of the numbers, there is only one possible method of numbering the boundaries around σ, and this is shown in Fig. 28a. Let us join the country σ_1 to σ; the new boundary MN dividing the countries $\sigma_1' = \sigma_1 + \sigma$ and $\sigma_2' = \sigma_2$ (Fig. 28b) will bear the number 1, while the numbering of the remaining boundaries remains unchanged. The map S', obtained in this way, will be normal, and its boundaries will be properly numbered with three digits. Since the number of countries in S' is n, it is possible to color it properly with four colors; if, for example, the country σ_1' has the color α, then the country σ_2' will be colored β. Restoring the country σ and coloring it γ, we obtain an admissible proper coloring of the map S with four colors.

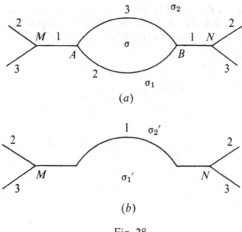

(a)

(b)

Fig. 28

Case 2. σ has *three* boundaries. The only possible way of numbering the boundaries around σ is shown in Fig. 29a. Let us imagine that the map S is drawn in a rubber film and that the country σ is constricted to a point; then the boundaries AB, BC, and AC will disappear, while the vertices A, B, and C will merge into one vertex $A = B = C = A'$ (Fig. 29b). The numbering of the boundaries MA', NA', and PA' (formerly MA, NB, PC), as well as of the other boundaries, remains the same, and we obtain a normal map S' with properly numbered boundaries. Since the number of countries in the map S' is n, it can be properly colored with four colors; if the country σ_1' is colored α, then the country σ_2' will be colored δ and the country σ_3' will be colored γ. Restoring the country σ and coloring it β, we obtain an admissible coloring of the map S.

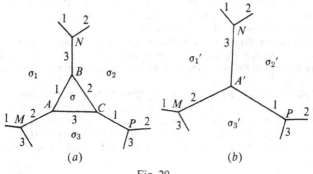

(a)   (b)

Fig. 29

Case 3. σ has *four* boundaries. In this case there are two essentially different possible ways of numbering the boundaries around the country σ (Fig. 30a and 31a). Let us examine the first way (Fig. 30a). It is always possible

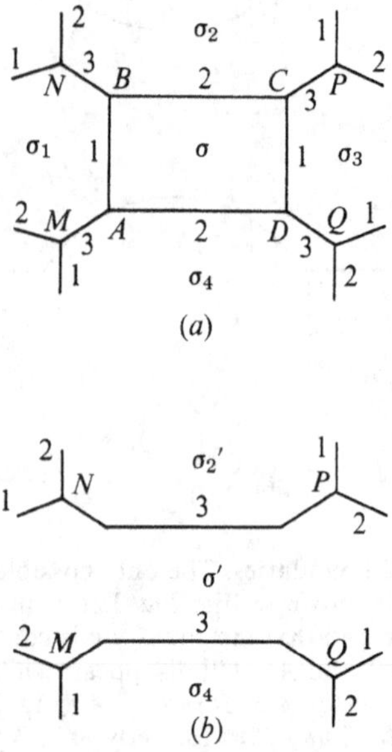

(a)

(b)

Fig. 30

to find two countries adjacent to σ which do not have a common boundary (see Example 16). Since both pairs of countries σ_1, σ_3 and σ_2, σ_4 which lie opposite each other are similar so far as the numbering of their boundaries is concerned, we may assume that in one of the pairs, say σ_1 and σ_3, the countries have no common boundary. Let us join both the countries σ_1 and σ_3 to σ, and assign the number 3 to new boundaries NP and MQ (Fig. 30b). The map S' obtained in this way will be normal, and its boundaries will be properly numbered. Since the number of countries in map S' is $n-1$, it is possible to color S' properly with four colors; if the country $\sigma' = \sigma_1 + \sigma_3 + \sigma$ is colored α, then the countries $\sigma_2' = \sigma_2$ and $\sigma_4' = \sigma_4$ will be colored δ. Restoring the country σ, we color it β, and again we obtain an admissible coloring for S.

With the second way of numbering (Fig. 31a), if the countries σ_1 and σ_3 do not have a common boundary, then it is possible to argue analogously. In this case, however, while the new boundary NP will be assigned the number 3 as before, the boundary MQ will be assigned the number 2 (Fig. 31b). The country $\sigma_4' = \sigma_4$ will be colored γ. Restoring the country σ, we color it β, as before.

Finally, let us suppose that the countries σ_2 and σ_4 do not have a common boundary. Let us draw the quadrilateral $ABCD$ together to a segment so that points A and B coincide, and points C and D coincide; thereupon the boundary BC will merge with the boundary AD. We number the boundaries MA, NB, PC, and QD just as before, and assign the number 1 to the new boundary $BC = AD$ (Fig. 31c). The map S' obtained in this way will be a normal map with properly numbered boundaries. Since the number of countries in the map S' is n, it is possible to color it properly with four colors; if the country σ_1' has the color α, then the country σ_2' will have the color δ, the country σ_3' the color α, and the country σ_4' the color γ. Restoring the country σ, we color it β.

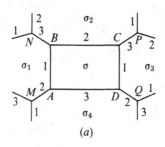

(a)

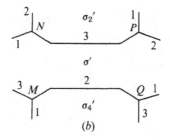

(b)

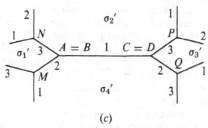

(c)

Fig. 31

Case 4. σ has *five* boundaries. In this case, except for permutations of the digits 1, 2, 3, there is only one possible way of numbering the boundaries around the country σ (Fig. 32a). Let us examine first the case in which

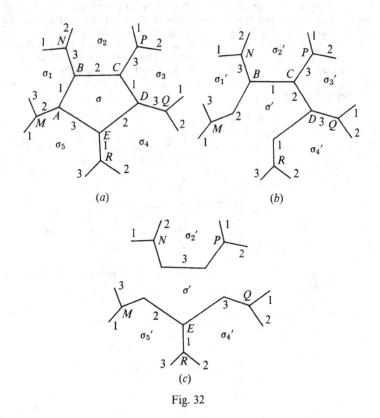

(a)

(b)

(c)

Fig. 32

country σ_5 neither coincides with σ_2 or σ_3 nor has boundaries in common with them. Let us join country σ_5 to σ, assign the number 2 to the new boundary MB, the number 1 to the new boundary RD, and change the number of the boundary BC to 1, and of the boundary CD to 2. In this way we obtain the normal map S' (Fig. 32b) with properly numbered boundaries. Since the number of countries in S' is n, it is possible to color it properly with four colors. If the country $\sigma' = \sigma + \sigma_5$ is colored α, then the countries σ_2' and σ_4' will be colored β, and the countries σ_1' and σ_3' will be colored γ. Restoring the country σ, we color it δ.

If country σ_5 adjoins or coincides with country σ_2, then the countries σ_1 and σ_3 neither have a common boundary nor coincide; if country σ_5 adjoins or coincides with σ_3, then σ_2 and σ_4 can neither coincide nor have a common boundary. Since both these cases are similar so far as numbering the boundaries is concerned, it is sufficient to consider the case when countries σ_1 and σ_3 neither coincide nor have a common boundary. Let us join both of these countries to σ, assign the number 3 to the new boundary NP, the number 2 to the new boundary ME, and the number 3 to the new boundary EQ. We obtain a normal map S' (Fig. 32c) with properly numbered boundaries. Since the number of countries in the map S' is $n - 1$, it is possible to color it properly with four colors. If the country $\sigma' = \sigma + \sigma_1 + \sigma_3$ is colored α, then the countries $\sigma_2' = \sigma_2$ and $\sigma_4' = \sigma_4$ will be colored δ, and the country $\sigma_5' = \sigma_5$ will be colored γ. Restoring the country σ, we color it β.

Just as it is not known whether every normal map can be properly colored with four colors, neither is it known whether the boundaries of every normal map can be properly numbered with three digits. We can prove only the following weaker assertion: The boundaries of any normal map can be properly numbered with four digits.

EXAMPLE 19. The boundaries of an arbitrary map (not even necessarily connected; see above p. 22) at each vertex of which *no more than* three boundaries meet can be properly numbered with four digits.

Proof. We shall carry out the proof by induction on n, the number of vertices of the map.

1. For $n = 2$ the assertion is obvious.

2. Let us assume that our assertion is valid for an arbitrary map of n vertices, at each of which no more than three boundaries meet, and let us consider a map S with $n + 1$ vertices which satisfies the same condition. Removing one of these vertices, say A_0, together with the boundaries which end at this vertex, we obtain a map S' with n vertices at each of which no more than three boundaries meet. By the inductive hypothesis, the boundaries of the map S' can be properly numbered with the four digits 1, 2, 3, and 4. Let us restore the vertex A_0 and its boundaries. Three possible cases arise:

Case 1. The vertex A_0 is connected (by means of one, two, or three boundaries) with only one vertex A_1 of the map S' (Fig. 33a, b, c). In this case a proper numbering of the boundaries of the map S' can easily be extended to a proper numbering of the boundaries of the map S.

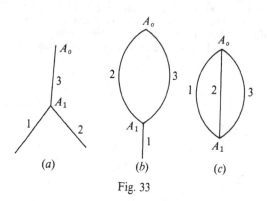

Fig. 33

Case 2. The vertex A_0 is connected with two vertices A_1 and A_2 of the map S', one of these perhaps being connected by two boundaries (Fig. 34a, b). It is easy to verify that in all cases a proper numbering of the boundaries of the map S' can be extended to a proper numbering of the boundaries of the map S.

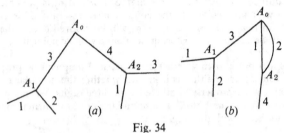

Fig. 34

Case 3. The vertex A_0 is connected with three vertices A_1, A_2, and A_3 of the map S' (Fig. 35). The least favorable case will be that in which two boundaries of the map S' pass through each of the vertices A_1, A_2, A_3. In this case, for each of the boundaries A_0A_1, A_0A_2, A_0A_3 we may select two possible numbers, for which three different numbers can be chosen only if the three pairs are identical, that is, if the three pairs of boundaries in the map S' issuing from the vertices A_1, A_2, and A_3 have the same numbers, for example, 1 and 2. In this case let us pick out on the map S' the longest curve which has its origin at the vertex A_1 and which consists of boundaries numbered alternately 1 and 3 (such a curve may consist of a single boundary and may end at either of the vertices A_2 or A_3). This curve cannot intersect itself because, by hypothesis, the boundaries of the map S' are properly numbered. Let us interchange the numbers 1 and 3 for the boundaries belonging to this curve. Obviously the boundaries of the map S' will remain properly numbered, and moreover with the new numbering, the three pairs of boundaries passing through the vertices A_1, A_2, and A_3 of the map S' will no longer be numbered alike; and in that case the proper numbering of the boundaries of the map S' can be easily extended to a proper numbering of the boundaries of the map S.

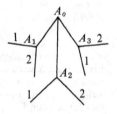

Fig. 35

3. Construction by Induction

We can apply the method of mathematical induction to the solution of construction problems if a positive integer n enters into the conditions of the problem (for example, the problems of constructing n-gons). In what follows we shall consider a number of examples of this type. In this chapter we shall include self-intersecting polygons (Fig. 36); in other words, by a polygon, in most problems, we mean *any* closed broken line $A_1 A_2 \ldots A_n$.

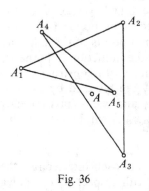

Fig. 36

14. CONSTRUCTING POLYGONS UNDER GIVEN CONDITIONS

EXAMPLE 20. Let $2n + 1$ points be given in the plane. Construct a $(2n + 1)$-gon for which these points are the mid-points of the sides.

SOLUTION. 1. For $n = 1$, the problem reduces to constructing a triangle given the mid-points of its sides, and the solution is easy. (It is sufficient to draw through each of the three given points a straight line parallel to the straight line joining the other two points.)

2. Let us assume that we can construct a $(2n - 1)$-gon given the mid-points of its sides, and let there be given $2n + 1$ points $A_1, A_2, \ldots, A_{2n+1}$ which are to be the mid-points of the sides of the desired $(2n + 1)$-gon $x_1 x_2 \ldots x_{2n+1}$.

Let us consider the quadrilateral $x_1 x_{2n-1} x_{2n} x_{2n+1}$ (Fig. 37). The points A_{2n-1}, A_{2n}, and A_{2n+1} are to be mid-points of the three sides $x_{2n-1} x_{2n}$, $x_{2n} x_{2n+1}$, and $x_{2n+1} x_1$. Let A be the mid-point of the fourth side $x_1 x_{2n-1}$. The quadrilateral $A_{2n-1} A_{2n} A_{2n+1} A$ will then be a parallelogram (for a proof it suffices to draw the straight line $x_1 x_{2n}$ and look at the triangles $x_1 x_{2n+1} x_{2n}$ and $x_1 x_{2n-1} x_{2n}$, in which the segments $A_{2n} A_{2n+1}$ and $A_{2n-1} A$ join the mid-points of the sides). Since the points A_{2n-1}, A_{2n}, and

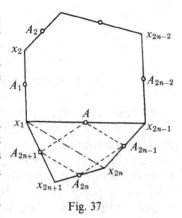

Fig. 37

A_{2n+1} are given, the fourth vertex A of the parallelogram can be easily constructed. This point A and the points $A_1, A_2, \ldots,$ A_{2n-2}, will then be mid-points of the sides of a $(2n - 1)$-gon $x_1 x_2 \ldots x_{2n-1}$, which, by the induction hypothesis, we are able to construct. All that remains is to construct the segments $x_1 x_{2n+1}$ and $x_{2n-1} x_{2n}$ so that they are bisected by the given points A_{2n+1} and A_{2n-1} (the points x_1 and x_{2n-1} have already been determined). Then the segment $x_{2n} x_{2n+1}$ will be bisected by the point A_{2n}.

When a polygon is not self-intersecting, the concept of exterior and interior points (with respect to the polygon) is clear. In the general case, however, this concept has no meaning; thus, for example, in Fig. 36 it is impossible to say whether the point A lies inside or outside the polygon. Instead we shall introduce the following definition. Let the arbitrary polygon $A_1 A_2 \ldots A_n$ be given. Let us establish a definite *direction of travel* around its vertices, say in the order $A_1, A_2, \ldots, A_n$. On one of the sides of the polygon, for example, $A_1 A_2$, let us construct a triangle $A_1 B A_2$. If the direction of travel around the vertices of the triangle in the order A_1, A_2, B is counter to the direction of travel around the vertices of the polygon (one clockwise and the other counterclockwise), then we shall say that the triangle *turns on the outside* of the polygon; if the directions of travel around the vertices of the triangle and of the polygon coincide, we shall say that the triangle *turns on the inside* of the polygon.

EXAMPLE 21. Given n points in the plane, construct an n-gon whose sides are the bases of isosceles triangles with apexes at the given n points and having the prescribed angles $\alpha_1, \alpha_2, \ldots, \alpha_n$ at the apexes.[1]

SOLUTION. We may allow some of the angles $\alpha_1, \alpha_2, \ldots, \alpha_n$ to be larger than 180°, with the understanding that for $\alpha < 180°$ the corresponding isosceles triangle turns on the outside of the polygon, and for $\alpha > 180°$ it turns on the inside (in this case the vertex angle of the triangle being $360° - \alpha$).

1. Let $n = 3$. Suppose that the problem has a solution and that x_1, x_2, x_3 are the vertices of the triangle we are seeking, and suppose that A_1, A_2, A_3 are the given apexes of the isosceles triangles constructed on its sides, with apex angles $\alpha_1, \alpha_2, \alpha_3$ (Fig. 38a). If

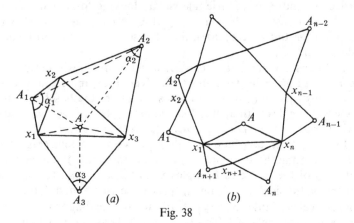

Fig. 38

we rotate the plane around the point A_1 through an angle α_1 (we agree that all rotations are counterclockwise), the vertex x_1 moves to the vertex x_2; if the rotation is about the point A_2 through an angle α_2, the vertex x_2 moves to the vertex x_3. These two rotations, performed successively, have the same effect as a single rotation through an angle $\alpha_1 + \alpha_2$ about a point A which can be constructed from the points A_1 and A_2 and the angles α_1 and α_2 in the following way: on the segment A_1A_2 we construct the angles $\dfrac{\alpha_1}{2}$ and $\dfrac{\alpha_2}{2}$ at the points A_1 and A_2, respectively; the point A of the intersection

[1] The preceding example may be regarded as a special case of Example 21 for which $\alpha_1 = \alpha_2 = \cdots = \alpha_n = 180°$.

of the sides of these angles (see Fig. 38a) will be the center of the resulting rotation through an angle $\alpha_1 + \alpha_2$.[1] Upon performing this resultant rotation the vertex x_1 moves to the vertex x_3. Consequently, the vertex x_3 will move to the vertex x_1 upon performing a rotation about the point A through an angle of $360° - (\alpha_1 + \alpha_2)$, and this means that the point A is the apex of an isosceles triangle with base x_1x_3 and apex angle $360° - (\alpha_1 + \alpha_2)$.

If the points A and A_3 do not coincide (this can happen only if $\alpha_1 + \alpha_2 + \alpha_3 = 360° \cdot k$), it is possible to construct the side x_1x_3.[2] To do this, we construct the angles $\dfrac{360° - (\alpha_1 + \alpha_2)}{2}$ and $\dfrac{\alpha_3}{2}$ on both sides of the segment AA_3 at the points A and A_3, respectively. The points of intersection of the sides of these angles will be the vertices x_1 and x_3 of the triangle we are looking for. Then it is not difficult to construct the vertex x_2. If $\alpha_1 + \alpha_2 + \alpha_3 = 360° \cdot k$ (if the points A and A_3 coincide), then the solution of the problem is indeterminate.

2. Let us assume that we can construct an n-gon whose sides are the bases of isosceles triangles with given apex angles, and let us try to construct an $(n + 1)$-gon whose sides are the bases of $n + 1$ isosceles triangles with apexes located at $A_1, A_2, \ldots, A_n, A_{n+1}$, and having apex angles $\alpha_1, \alpha_2, \ldots, \alpha_n, \alpha_{n+1}$, respectively.

Let $x_1x_2 \ldots x_nx_{n+1}$ be the $(n + 1)$-gon we are seeking (Fig. 38b). Let us examine the triangle $x_1x_nx_{n+1}$. As in part 1, the apexes A_n and A_{n+1} of the isosceles triangles $x_nA_nx_{n+1}$ and $x_{n+1}A_{n+1}x_1$ constructed on the sides x_nx_{n+1} and $x_{n+1}x_1$ can be used to find A, the apex of an isosceles triangle x_1Ax_n constructed on the diagonal x_1x_n and having an apex angle of $360° - (\alpha_n + \alpha_{n+1})$. Thus, our problem is reduced to the problem of constructing an n-gon $x_1x_2 \ldots x_n$ for which the points $A_1, A_2, \ldots, A_{n-1}, A$ are apexes of isosceles triangles constructed as its sides with the given apex angles $\alpha_1, \alpha_2, \ldots, \alpha_{n-1}, 360° - (\alpha_n + \alpha_{n+1})$. By virtue of the induction hypothesis, the n-gon $x_1x_2 \ldots x_n$ may be constructed, and it is then easy to construct the desired $(n + 1)$-gon $x_1x_2 \ldots x_nx_{n+1}$ as well.

If $\alpha_1 + \alpha_2 + \cdots + \alpha_{n+1} = 360° \cdot k$, the solution of the problem is either impossible or indeterminate (why?).

---

[1] See, for example, Part I, Chapter I, section 2 of I. M. Yaglom, *Geometric Transformations,* translated by A. Shields. Part I is No. 8 of the New Mathematical Library (Syracuse, N. Y.: The L. W. Singer Company, Inc.; New York: Random House, Inc.; 1962).

[2] Here k is some integer.

Problem 14. Given n points in the plane, construct an n-gon for which these points are vertices of triangles with bases the sides of the n-gon, and for which these triangles have given vertex angles and given ratios for the lateral sides.

Hint. This problem may be solved in a manner analogous to that of the preceding exercise (the preceding exercise is a special case of this one), only in place of a rotation around a given point A_1 through a given angle α_1, we should consider a similarity transformation consisting of a rotation through an angle α_1 and a central similitude (homothetic transformation) with the same point A_1 as the center and with the coefficient of similarity equal to the ratio of the sides of the corresponding triangle (and analogously for the other given points).[1] The result of two such transformations is equivalent to a third transformation of the same type.[2] Consequently, using a notation analogous to that used in the solution of the preceding example, it is possible to use the vertices A_1 and A_2 of the triangles $x_1x_2A_1$ and $x_2x_3A_2$ to find A, the vertex of the triangle x_1x_3A, constructed on the segment x_1x_3 and having a given angle as vertex angle and a given ratio as the ratio of its lateral sides.

The construction of the sides x_1x_3 of the triangle $x_1x_2x_3$ from the points A and A_3 may be carried out, for example, in the following way: The two similarity transformations with centers at A and A_3 carried out one after the other, will transform x_1 into itself (first x_1 moves to x_3 and then x_3 moves to x_1). But this sequence of transformations is equivalent to a single similarity transformation with center at a point B which it is possible to construct. Since the point B is transformed into itself, then it must coincide with the point x_1 which we are seeking. If the sum of the given vertex angles is a multiple of $360°$, and the product of the ratios of the sides equals one, then the solution of the problem is either impossible or indeterminate.

EXAMPLE 22. Let a circle and n points be given in the plane. In the circle inscribe an n-gon whose sides pass through the given points.

SOLUTION. This problem is difficult. For the solution, the method of mathematical induction has to be used in a completely unexpected way. That is, this time we are unable to carry out the induction on n, the number of sides of the polygon; instead we have to look at the more general problem of constructing an n-gon for which k adjacent sides go through k given points, while the re-

---

[1] A transformation consisting of an enlargement or reduction in which the distance of every point from a fixed point O is changed by a fixed ratio ρ (in the figure, the transformation carries A into A', and $OA' = \rho \cdot OA$) is called a *central similitude* with center O and coefficient of similarity ρ.

[2] See, for example, Part II, Chapter I, section 2 of the book by I. M. Yaglom noted on p. 48. Part II is to appear as a later number of the New Mathematical Library.

maining $n - k$ sides are parallel to given straight lines, and carry out the induction on k (this problem reduces to the given one for $k = n$).

1. For $k = 1$ we have the following problem: In the given circle, inscribe an n-gon for which the side A_1A_n passes through a given point P, while the remaining $n - 1$ sides A_1A_2, A_2A_3, ..., $A_{n-1}A_n$ are parallel to given straight lines $l_1, l_2, ..., l_{n-1}$.

Let us suppose that the problem has a solution and that the polygon $A_1A_2 ... A_n$ we seek is constructed (Fig. 39a, b). On the circle,

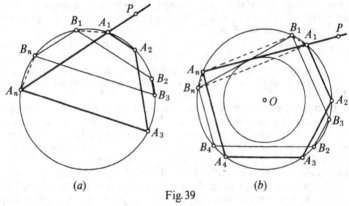

(a)   (b)

Fig. 39

let us take an arbitrary point B_1 and construct an inscribed polygon $B_1B_2 ... B_n$ whose sides B_1B_2, B_2B_3, ..., $B_{n-1}B_n$ are parallel to the straight lines $l_1, l_2, ..., l_{n-1}$, respectively. Then the arcs A_1B_1, A_2B_2, ..., A_nB_n will be equal, the pairs of arcs A_1B_1 and A_2B_2, A_2B_2 and A_3B_3, etc., being oppositely directed on the circle. Consequently, for even n the arcs A_1B_1 and A_nB_n will be oppositely directed and the quadrilateral $A_1B_1B_nA_n$ will be an isosceles trapezoid with the bases A_1A_n and B_1B_n (Fig. 39a). Hence, the side A_1A_n of the polygon we are seeking is parallel to the side B_1B_n of the n-gon $B_1B_2 ... B_n$; consequently, in this case it is necessary to draw a straight line through the point P parallel to B_1B_n, after which it is not difficult to determine the remaining vertices of the n-gon $A_1A_2 ... A_n$. (Carry out the investigation.)[1]

[1] It is conceivable that a line through P parallel to B_1B_n might not intersect the circle. More generally, in a construction problem the reader must investigate every step to determine exactly what conditions must be placed on the given data in order that the step can be performed. For many steps the data can be given so that the step is impossible to perform. In this and the following examples and exercises this investigation is left to the reader.

For odd n, the arcs A_1B_1 and A_nB_n will have the same direction on the circle and the quadrilateral $A_1B_1A_nB_n$ will be an isosceles trapezoid with bases A_1B_n and B_1A_n (Fig. 39b). Since the diagonals A_1A_n and B_1B_n of the trapezoid are equal, in this case it is necessary to draw through the point P a straight line on which the given circle cuts off a chord A_1A_n equal to the known chord B_1B_n, that is, a straight line tangent to the circle concentric with the given one and tangent to B_1B_n (look at this!).

2. Let us assume that we can already solve the problem of inscribing in a circle an n-gon of which k successive sides pass through k given points, while the remaining $n - k$ sides are parallel to given straight lines, and let us attempt to inscribe in a circle an n-gon for which $k + 1$ adjacent sides A_1A_2, A_2A_3, . . . , $A_{k+1}A_{k+2}$ pass through $k + 1$ given points P_1, P_2, . . . , P_{k+1}, while the remaining $n - k - 1$ sides are parallel to given straight lines.

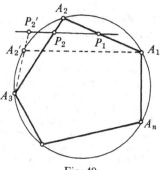

Fig. 40

Let us assume that the problem has been solved and that the desired n-gon ($k + 1$ sides through $k + 1$ points) has been constructed (Fig. 40). Let us consider the sides A_1A_2 and A_2A_3 of this polygon. Let us draw a straight line A_1A_2' through the vertex A_1 parallel to P_1P_2, and let A_2' be the point of intersection of this straight line with the circle, and P_2' be the point of intersection of the line $A_2'A_3$ with P_1P_2. The triangles $P_1A_2P_2$ and $P_2'P_2A_3$ are similar, for

$$\angle A_2P_1P_2 = \angle A_2A_1A_2' = \angle A_2A_3P_2'$$

and

$$\angle A_2P_2P_1 = \angle P_2'P_2A_3.$$

Consequently,

$$\frac{P_1P_2}{A_3P_2} = \frac{A_2P_2}{P_2'P_2},$$

from which we have

$$P_2'P_2 = \frac{A_3P_2 \cdot A_2P_2}{P_1P_2}.$$

Since the product $A_3P_2 \cdot A_2P_2$ depends only on the given point P_2 and the circle (but not on the choice of the points A_2 and A_3 even if P_2 is on or outside the circle!) the product may be determined; therefore, the magnitude of the segment $P_2'P_2$ may be found and, consequently, the point P_2' may be constructed. In this way we have k known points $P_2', P_3, \ldots, P_{k+1}$, through which pass the k adjacent sides $A_2'A_3, A_3A_4, \ldots, A_{k+1}A_{k+2}$ of the n-gon $A_1A_2'A_3 \ldots A_n$, while the remaining $n - k$ sides $A_{k+2}A_{k+3}, \ldots, A_nA_1, A_1A_2'$ are parallel to known straight lines. By the inductive hypothesis we can construct the n-gon $A_1A_2'A_3 \ldots A_n$, after which it is easy to construct the desired n-gon $A_1A_2 \ldots A_n$ as well.

Problem 15. In a given circle inscribe an n-gon, k sides (not necessarily adjacent!) of which pass through k given points, while the remaining $n - k$ sides are parallel to given straight lines.

Hint. Let the side A_1A_2 of the desired polygon pass through the point P and let the side A_2A_3 be parallel to the straight line l (Fig. 41). Let P' denote the point symmetric to P with respect to the diameter of the circle which is perpendicular to the straight line l, and let A_2' be the point of intersection of the straight line $P'A_3$ with the circle. In the n-gon $A_1A_2'A_3 \ldots A_n$ the side A_1A_2' is parallel to the given straight line l, while the side $A_2'A_3$ passes through the known point P'. Carrying out this construction a suitable number of times we can reduce this problem to that of the construction of an n-gon of which k *adjacent* sides pass through the given points while the remaining $n - k$ sides are parallel to the given straight lines.

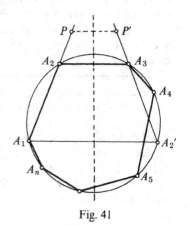

Fig. 41

15. DIVISION OF A LINE SEGMENT

EXAMPLE 23. Given two parallel straight lines l and l_1, using a straightedge alone, divide a segment AB on the straight line l into n equal parts.

SOLUTION. 1. Let $n = 2$. Let S be an arbitrary point of the plane not lying on the straight lines l and l_1, and join S with the points A and B (Fig. 42a). Let C and D be the points of intersec-

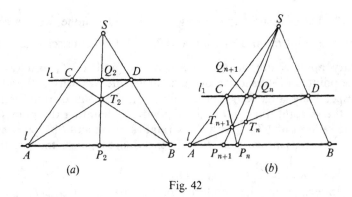

Fig. 42

tion of the straight lines AS and BS with the straight line l_1. Let T_2 be the intersection point of the straight lines AD and BC, and let P_2 be the point of intersection of the straight lines ST_2 and l. Let us prove that P_2 is the desired point, that is, that $AP_2 = \frac{1}{2}AB$.

Let Q_2 denote the point of intersection of the straight lines ST_2 and l_1. It is easy to see that

$$\triangle T_2P_2B \sim \triangle T_2Q_2C,$$
$$\triangle ABT_2 \sim \triangle DCT_2,$$
$$\triangle SAP_2 \sim \triangle SCQ_2,$$

and

$$\triangle SAB \sim \triangle SCD,$$

from which we have

$$\frac{P_2B}{Q_2C} = \frac{T_2B}{T_2C} = \frac{AB}{CD}$$

and

$$\frac{P_2A}{Q_2C} = \frac{SA}{SC} = \frac{AB}{CD}.$$

Consequently,

$$\frac{P_2B}{Q_2C} = \frac{P_2A}{Q_2C},$$

and therefore $P_2A = P_2B$ and $AP_2 = \frac{1}{2}AB$.

2. Let us assume that, using a straightedge alone, we can construct a point P_n on AB such that $AP_n = \frac{1}{n} AB$. Let us select an arbitrary point S not on the lines l and l_1 and let T_n and Q_n be the points of intersection of the straight line SP_n with AD and l_1, respectively (Fig. 42b). Let us join T_{n+1}, the point of intersection of the straight lines AD and CP_n, with S, denoting the points of intersection of ST_{n+1} with the straight lines l_1 and l by Q_{n+1} and P_{n+1}. Let us prove that P_{n+1} is the desired point, that is, that $AP_{n+1} = \frac{1}{n+1} AB$.

Indeed, from the similarity of the triangles $CQ_{n+1}T_{n+1}$ and $P_nP_{n+1}T_{n+1}$, and triangles $CT_{n+1}D$ and $P_nT_{n+1}A$, we have

$$\frac{P_{n+1}P_n}{CQ_{n+1}} = \frac{P_nT_{n+1}}{CT_{n+1}} = \frac{AP_n}{CD}; \tag{11}$$

from the similarity of the triangles SAP_{n+1} and SCQ_{n+1}, and triangles SAB and SCD, we have

$$\frac{AP_{n+1}}{CQ_{n+1}} = \frac{SA}{SC} = \frac{AB}{CD}. \tag{12}$$

From equalities (11) and (12) it follows that

$$\frac{P_{n+1}P_n}{AP_{n+1}} = \frac{AP_n}{AB}$$

or, since $P_{n+1}P_n = AP_n - AP_{n+1}$ and $AP_n = \frac{1}{n} AB$,

$$\frac{\frac{1}{n} AB - AP_{n+1}}{AP_{n+1}} = \frac{\frac{1}{n} AB}{AB},$$

$$\frac{1}{n} AB - AP_{n+1} = \frac{1}{n} AP_{n+1},$$

and, finally,

$$AP_{n+1} = \frac{1}{n+1} AB.$$

To find the points P_{n+1}', P_{n+1}'', ... of division, we can use the same method to construct the segments P_{n+1}, $P_{n+1}' = \frac{1}{n} P_{n+1}B$,

$P_{n+1}'P_{n+1}'' = \frac{1}{n-1} P'B$, etc.

Problem 16. Using a compass with a fixed spread a, and a straight-edge, construct a line segment of length $\frac{1}{n} a$.

Hint. On a circle of radius a, locate points A_1, A_2, A_3, A_4, A_5, A_6 which are vertices of a regular hexagon. Let us assume that we can locate a point B_n on the radius OA_n such that $OB_n = \frac{1}{n} OA_n = \frac{a}{n}$. (Here we consider $A_{6m+k} = A_k$ for arbitrary m, and $k = 1, 2, 3, 4, 5, 6$; $B_1 = A_1$.) Let B_{n+1} be the point of intersection of the straight lines OA_{n+1} and $B_n A_{n+2}$; then $OB_{n+1} = \frac{a}{n+1}$.

4. The Determination of Geometric Loci by Induction

In this chapter we shall consider several problems concerning the determination of geometric loci by the method of induction.

16. PROBLEMS INVOLVING SUMS OF AREAS

Let $S(ABC)$ denote the area of the triangle ABC.

EXAMPLE 24. Segments $B_1C_1, B_2C_2, \ldots, B_nC_n$ are laid off on the sides of a convex n-gon $A_1A_2 \ldots A_n$. Find the locus of all of the points M interior to the polygon for which the sum of the areas of the triangles $MB_1C_1, MB_2C_2, \ldots, MB_nC_n$ is a constant (equal to the sum $S(M_0B_1C_1) + S(M_0B_2C_2) + \cdots + S(M_0B_nC_n)$, where M_0 is some fixed point interior to the polygon).

SOLUTION. 1. Let $n = 3$ (Fig. 43a). On the sides A_3A_2 and A_3A_1 of the triangle $A_1A_2A_3$ let us lay off segments $A_3P = B_2C_2$

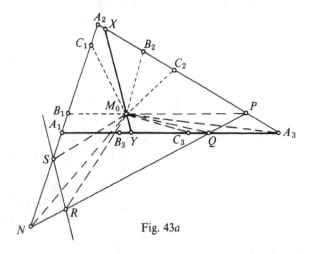

Fig. 43a

and $A_3Q = B_3C_3$. Then[1]

$$S(M_0B_2C_2) + S(M_0B_3C_3) = S(M_0PA_3) + S(M_0QA_3)$$
$$= S(PQA_3) + S(M_0PQ),$$

and, consequently,

$$S(M_0B_1C_1) + S(M_0B_2C_2) + S(M_0B_3C_3)$$
$$= S(PQA_3) + [S(M_0B_1C_1) + S(M_0PQ)].$$

Analogously,

$$S(MB_1C_1) + S(MB_2C_2) + S(MB_3C_3)$$
$$= S(PQA_3) + [S(MB_1C_1) + S(MPQ)].$$

We see that the desired locus is determined by the condition that

$$S(MB_1C_1) + S(MPQ) = S(M_0B_1C_1) + S(M_0PQ).$$

Now let N be the point of intersection of the straight lines A_1A_2 and PQ. (If these straight lines are parallel, then the locus we are seeking is a segment of a straight line parallel to these lines.) On the sides of the angle A_2NP let us lay off segments $NR = PQ$ and $NS = B_1C_1$; then

$$S(M_0B_1C_1) + S(M_0PQ) = S(M_0NS) + S(M_0NR)$$
$$= S(NRS) + S(M_0RS),$$

and, analogously,

$$S(MB_1C_1) + S(MPQ) = S(NRS) + S(MRS).$$

Consequently, the locus we are seeking consists of those points M lying inside the triangle for which $S(MRS) = S(M_0RS)$; that is, it is the segment XY of the straight line passing through the point M_0 and parallel to the straight line RS.[2]

2. Suppose we already know that the desired locus for the n-gon is a straight-line segment (passing, of course, through the point M_0). Let us now consider an $(n + 1)$-gon $A_1A_2 \ldots A_nA_{n+1}$; let $B_1C_1, B_2C_2, \ldots, B_nC_n, B_{n+1}C_{n+1}$ be the given segments laid off on

[1] Here we assume that the point M_0 lies inside the quadrilateral A_1A_2PQ; the reasoning would be but a little different if it were otherwise.

[2] We could begin the induction with $n = 2$, in which case the "n-gon" would be just an angle with only two sides.

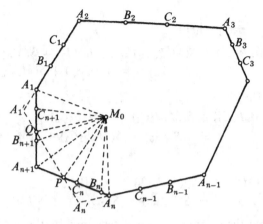

Fig. 43b

its sides, and let M_0 be the point inside the $(n + 1)$-gon (Fig. 43b). On the sides of the angle $A_1 A_{n+1} A_n$ from the vertex A_{n+1} let us lay off segments $A_{n+1}P = B_n C_n$ and $A_{n+1}Q = B_{n+1} C_{n+1}$. Then

$$S(MB_n C_n) + S(MB_{n+1} C_{n+1}) = S(MA_{n+1}P) + S(MA_{n+1}Q)$$
$$= S(A_{n+1}PQ) + S(MPQ).$$

Consequently, for the points M of the locus we are seeking

$$S(MB_1 C_1) + S(MB_2 C_2) + \cdots + S(MB_{n-1} C_{n-1}) + S(MPQ)$$
$$= S(M_0 B_1 C_1) + S(M_0 B_2 C_2) + \cdots + S(M_0 B_{n-1} C_{n-1}) + S(M_0 PQ).$$

Thus, by the inductive hypothesis, the desired locus is a straight line segment which passes through the point M_0.

From our method of solving this problem, it is easy to discover a method for constructing this locus.

Problem 17. Let n straight lines $l_1, l_2, \ldots, l_n$ be given, on each of which is given a line segment $B_1 C_1, B_2 C_2, \ldots, B_n C_n$, and let M_0 be a fixed point. Find the locus of points M such that the algebraic sum of the areas of the triangles $MB_1 C_1, MB_2 C_2, \ldots, MB_n C_n$ is equal to the corresponding sum for the point M_0. Here the area of the triangle $MB_i C_i$ $(i = 1, 2, \ldots, n)$ is positive if M lies on the same side of the straight line l_i as the point M_0, and negative in the opposite case.

Hint. The desired locus is a straight line; the proof is analogous to the solution in Example 24.

Problem 18. Prove that if a quadrilateral is circumscribed about a circle, then the mid-points of the diagonals and the center of the circle lie on a straight line (Fig. 44).

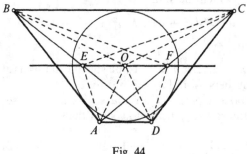

Fig. 44

Hint. Using the notation of Fig. 44 we have

$$S(BCE) + S(ADE) = S(BCF) + S(ADF)$$

$$= S(BCO) + S(ADO) = \frac{1}{2} S,$$

where S is the area of the quadrilateral. From this, by virtue of the results of Example 24 (or Problem 17), it follows that the points E, F, and O lie on a straight line.

Problem 19. Prove that the straight line joining the mid-points of the diagonals of a convex quadrilateral (not a parallelogram and not a trapezoid) bisects the line segment which joins the points of intersection of the pairs of opposite sides (Fig. 45).

Hint. Using the notation of Fig. 45 (where P is the mid-point of the segment EF), we have

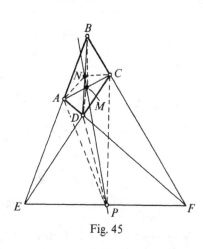

Fig. 45

$$S(ABM) + S(CDM) = S(ABN) + S(CDN) = S(ABP) - S(CDP) = \frac{1}{2} S,$$

where S is the area of the quadrilateral. Hence, from the result of Problem 17, it follows that the points M, N, and P lie on a straight line.

17. PROBLEMS INVOLVING SQUARES OF DISTANCES

EXAMPLE 25. Given n points $A_1, A_2, \ldots, A_n$ and n numbers $a_1, a_2, \ldots, a_n$ (positive or negative!), find the geometric locus of the points M for which the sum

$$a_1 \cdot MA_1{}^2 + a_2 \cdot MA_2{}^2 + \cdots + a_n \cdot MA_n{}^2$$

is a constant.

SOLUTION. 1. Let $n = 2$. To be specific, let us assume first that both the numbers a_1 and a_2 are positive.

Let us take a point O on the segment $A_1 A_2$ such that it is divided in the ratio a_2/a_1, that is, such that

$$A_1 O = \frac{a_2}{a_1 + a_2} A_1 A_2 \quad \text{and} \quad O A_2 = \frac{a_1}{a_1 + a_2} A_1 A_2.$$

Let M be an arbitrary point of the plane, and let H be the foot of the perpendicular dropped from M to the straight line $A_1 A_2$ (Fig. 46). Then we have

$$MA_1{}^2 = MO^2 + A_1 O^2 \pm 2A_1 O \cdot OH,$$
$$MA_2{}^2 = MO^2 + O A_2{}^2 \mp 2O A_2 \cdot OH.$$

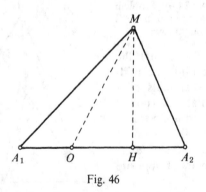

Fig. 46

Multiplying the first of these equalities by OA_2 and the second by $A_1 O$ and adding the resulting equalities termwise, we obtain

$$MA_1{}^2 \cdot A_2 O + MA_2{}^2 \cdot A_1 O$$
$$= MO^2(OA_2 + A_1 O) + A_1 O^2 \cdot OA_2 + OA_2{}^2 \cdot A_1 O$$
$$= MO^2 \cdot A_1 A_2 + A_1 O \cdot OA_2 \cdot A_1 A_2.$$

Now let us substitute the values of A_1O and OA_2; we obtain

$$MA_1{}^2 \frac{a_1 \cdot A_1A_2}{a_1 + a_2} + MA_2{}^2 \frac{a_2 \cdot A_1A_2}{a_1 + a_2}$$

$$= MO^2 \cdot A_1A_2 + \frac{a_1}{a_1 + a_2} \cdot \frac{a_2}{a_1 + a_2} A_1A_2{}^3,$$

or

$$a_1 MA_1{}^2 + a_2 MA_2{}^2 = (a_1 + a_2) MO^2 + \frac{a_1a_2}{a_1 + a_2} A_1A_2{}^2.$$

Hence, if $a_1 MA_1{}^2 + a_2 MA_2{}^2 = R^2$, a constant by hypothesis, then

$$MO^2 = \frac{R^2}{a_1 + a_2} - \frac{a_1a_2}{(a_1 + a_2)^2} A_1A_2{}^2 = \text{constant}.$$

From this it follows that if

$$K \equiv \frac{R^2}{a_1 + a_2} - \frac{a_1a_2}{(a_1 + a_2)^2} A_1A_2{}^2 > 0,$$

then the desired locus of the points M will be a circle with center at the point O and radius $\sqrt{K}$; if $K = 0$, then the desired locus is the single point O; finally, if $K < 0$, then this locus does not have any points.

The case in which a_1 and a_2 are both negative can be reduced to the preceding case in an obvious way. If $a_1 > 0$, $a_2 < 0$, and $a_1 + a_2 \neq 0$ (for example if $a_1 + a_2 > 0$), the point O should be taken on the extension of the segment A_1A_2 to the right of the point A_2 so that

$$A_2O = \left| \frac{a_1}{a_1 + a_2} \right| \quad \text{and} \quad A_1O = \left| \frac{a_2}{a_1 + a_2} \right|.$$

The remainder of the argument will not differ from that given above.

Finally, if $a_1 + a_2 = 0$, then $a_1 = -a_2$ and our problem is reduced to the following: Find the locus of the points M for which the difference of the squares of the distances from two given points A_1 and A_2 is a constant. Let H be the foot of the perpendicular dropped from the point M to the straight line A_1A_2 (Fig. 46); then $MA_1{}^2 = MH^2 + A_1H^2$, $MA_2{}^2 = MH^2 + HA_2{}^2$, and, consequently, $MA_1{}^2 - MA_2{}^2 = A_1H^2 - HA_2{}^2$. If $MA_1{}^2 - MA_2{}^2 = R^2$, then $A_1H - HA_2 = R^2/A_1A_2$, which completely determines the point H; from this it follows that the desired locus in this case is a straight line passing through the point H and perpendicular to A_1A_2.

2. Let us assume that we have proved that for n given points the corresponding locus is a circle if $a_1 + a_2 + \cdots + a_n \neq 0$ and a straight line if $a_1 + a_2 + \cdots + a_n = 0$. Let us now consider $n + 1$ points $A_1, A_2, \ldots, A_{n+1}$ and $n + 1$ numbers $a_1, a_2, \ldots, a_{n+1}$. Let us assume that $a_n + a_{n+1} \neq 0$. (If $a_n + a_{n+1} = 0$, then we would substitute the numbers a_{n-1} and a_{n+1}, or a_{n-1} and a_n for this pair of numbers; and if it happened that simultaneously $a_n + a_{n+1} = 0$, $a_{n-1} + a_{n+1} = 0$, and $a_{n-1} + a_n = 0$, then $a_{n-1} = a_n = a_{n+1} = 0$, and we could use the inductive hypothesis directly, because the problem would then reduce to the case of $n - 2$ points $A_1, A_2, \ldots, A_{n-2}$ and $n - 2$ numbers $a_1, a_2, \ldots, a_{n-2}$.)

As in part 1, we show that it is possible to find a point O on the segment $A_n A_{n+1}$ such that for an arbitrary point M of the plane,

$$a_n MA_n{}^2 + a_{n+1} MA_{n+1}{}^2 = (a_n + a_{n+1})MO^2 + \frac{a_n a_{n+1}}{a_n + a_{n+1}} A_n A_{n+1}{}^2.$$

In the same way our problem is reduced to finding the locus of points M for which the sum

$$a_1 MA_1{}^2 + a_2 MA_2{}^2 + \cdots + a_{n-1} MA_{n-1}{}^2 + (a_n + a_{n+1})MO^2$$

is a constant. By the inductive hypothesis this locus will be a circle if $a_1 + a_2 + \cdots + a_n + a_{n+1} \neq 0$ and a straight line if $a_1 + a_2 + \cdots + a_n + a_{n+1} = 0$.

Problem 20. Find the locus of the points for which the sum of the squares of the distances from n given points is a constant.

Hint. It suffices to put $a_1 = a_2 = \cdots = a_n = 1$ in Example 25.

Problem 21. Find the point for which the sum of the squares of the distances from n given points is a minimum.

Hint. Consider the center of the circle which is the desired locus in Problem 20.

Problem 22. Find the locus of the points for which the ratio of the distances from two given points is a constant.

Hint. If M is a point of the desired locus, then $AM/BM = c$, and, consequently, $AM^2 - c^2 BM^2 = 0$; hence, this problem is reduced to Example 25.

Problem 23. Let the n-gon $A_1A_2 \ldots A_n$ be given. Find the locus of the points M such that the polygon whose vertices are the projections of M on the sides of the given polygon have a given area S.

Hint. It is possible to show that the area of a triangle whose vertices are the projections of M on the sides of the triangle $A_1A_2A_3$ is given by

$$\frac{1}{4} \left| 1 - \frac{d^2}{R^2} \right| S(A_1A_2A_3),$$

where R is the radius of the circumscribed circle Σ of the triangle $A_1A_2A_3$ and d is the distance of the point M from the center of the circle Σ. From this it follows that for $n = 3$ the desired locus is a circle concentric with Σ (or a pair of such circles). Further, using induction on the number of sides of the polygon, it can be shown that for arbitrary n, the desired locus will be a circle (or a pair of concentric circles).

5. Definitions by Induction

Interesting examples of applications of the method of mathematical induction in geometry are found in problems containing concepts whose very definitions involve the process of going "from n to $n + 1$." This chapter is devoted to problems of this type.

18. CENTER OF GRAVITY AND MEDIANS OF POLYGONS

EXAMPLE 26. Define the medians and the center of gravity of an n-gon.

SOLUTION. 1. The *center of gravity of a line segment* will be defined as its mid-point (Fig. 47a). (In this chapter we shall consider a line segment as being a 2-gon.)

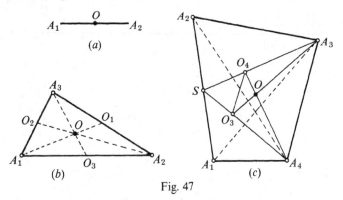

Fig. 47

Then the *medians of a triangle* $A_1A_2A_3$ may be defined as the line segments which join vertices of the triangle to the centers of gravity of the opposite sides (Fig. 47b). As is well known, the medians of a triangle intersect in a common point, and are divided at that point in the ratio 2:1, with the longer portion toward the vertex. The point of intersection of all of the medians of a triangle is called the *center of gravity of the triangle*.

Now let us agree to call those line segments which join vertices of the quadrilateral $A_1A_2A_3A_4$ with the centers of gravity O_1, O_2, O_3, O_4 of the triangles composed of the three remaining vertices, the *medians of the quadrilateral* (Fig. 47c). We shall prove that *the medians of a quadrilateral intersect in a common point and that this point divides each of them in the ratio 3:1, with the longer portion toward the vertex*. Indeed, let S denote the center of gravity (midpoint) of the side A_1A_2; let O_4 and O_3 denote the centers of gravity of the triangles $A_1A_2A_3$ and $A_1A_2A_4$, respectively; and, further, let O be the point of intersection of the medians A_3O_3 and A_4O_4 of the quadrilateral. Since SA_3 and SA_4 are medians of the triangles $A_1A_2A_3$ and $A_1A_2A_4$, respectively, then

$$\frac{SA_3}{SO_4} = \frac{3}{1}$$

and

$$\frac{SA_4}{SO_3} = \frac{3}{1},$$

and, consequently,

$$\frac{SA_3}{SO_4} = \frac{SA_4}{SO_3}.$$

From this it follows that O_3O_4 is parallel to A_3A_4 and that

$$\frac{A_3A_4}{O_3O_4} = \frac{SA_3}{SO_4} = \frac{3}{1}.$$

Further, from the similarity of triangles OO_3O_4 and OA_3A_4 we have

$$\frac{OA_4}{OO_4} = \frac{OA_3}{OO_3} = \frac{A_3A_4}{O_3O_4} = \frac{3}{1}.$$

Thus, an arbitrary pair of adjacent medians of a quadrilateral (that is, medians issuing from adjacent vertices) are divided at their point of intersection in the ratio 3:1. From this it follows that all four of the medians of the quadrilateral pass through a common point O, and this point divides each in the ratio 3:1. The point of intersection O of the medians of a quadrilateral is called the *center of gravity of the quadrilateral*.[1]

[1] For a general quadrilateral, this point will not coincide with the physical center of gravity, which by definition is the point of balance of a thin homogeneous metal plate having the shape of the given quadrilateral. A different name might have been selected for the point defined in this section. Actually the point we determine by the method of Example 26 is the physical center of gravity of a system of n unit weights, one placed at each vertex of the n-gon. See, for example, pp. 11–14 of *Geometry* by B. V. Kutuzov (School Mathematics Study Group, Studies in Mathematics, vol. IV, distributed by A. C. Vroman, Inc., Pasadena, California).

2. Let us assume that for all $k < n$ we have defined the medians of a k-gon as the line segments joining vertices of the k-gon with the centers of gravity of the $(k-1)$-gons formed from the remaining $k - 1$ vertices, and that for all $k < n$ we have defined the center of gravity of a k-gon as the point of intersection of its medians. We shall further assume that it has been proved that for $k < n$ the medians of the k-gon are divided by their point of intersection (the center of gravity of the k-gon) in the ratio $(k - 1):1$ (with the longer side nearer the vertex).

Let us now define the *medians of an n-gon* as the line segments joining vertices of the n-gon with the centers of gravity of the $(n-1)$-gons formed from the remaining $n - 1$ vertices. We shall prove that *all of the medians of an n-gon $A_1A_2 \ldots A_n$ intersect in a common point, which divides each of them in the ratio $(n-1):1$* (with the longer portion toward the vertex). Indeed, let S be the center of gravity of the $(n-2)$-gon $A_1A_2 \ldots A_{n-2}$; then the straight lines SA_{n-1} and SA_n will be medians of the $(n-1)$-gons $A_1A_2 \ldots A_{n-1}$ and $A_1A_2 \ldots A_{n-2}A_n$ (Fig. 48). If O_n and O_{n-1} are the centers of

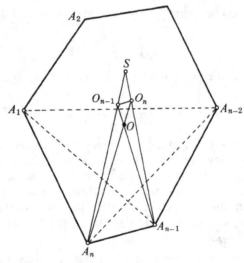

Fig. 48

gravity of these $(n-1)$-gons, then, by the inductive hypothesis,

$$\frac{SA_{n-1}}{SO_n} = \frac{SA_n}{SO_{n-1}} = \frac{n-1}{1}.$$

Consequently, $O_{n-1}O_n$ is parallel to A_nA_{n-1} and

$$\frac{A_{n-1}A_n}{O_{n-1}O_n} = \frac{n-1}{1}.$$

Let O be the point of intersection of the medians $O_{n-1}A_{n-1}$ and O_nA_n of the n-gon $A_1A_2 \ldots A_n$. From the similarity of the triangles $OO_{n-1}O_n$ and $OA_{n-1}A_n$ it follows that

$$\frac{OA_{n-1}}{OO_{n-1}} = \frac{OA_n}{OO_n} = \frac{A_{n-1}A_n}{O_{n-1}O_n} = \frac{n-1}{1}.$$

Thus, an arbitrary pair of adjacent medians of an n-gon is divided by their point of intersection in the ratio $(n-1):1$. From this it follows that all of the medians of an n-gon intersect at a common point, and are divided by that point in the ratio $(n-1):1$.

Now we can define the *center of gravity of an n-gon* as the point of intersection of its medians, and then we can define the *medians of an $(n+1)$-gon* as the line segments joining vertices of the $(n+1)$-gon with the centers of gravity of the n-gons formed from the remaining n vertices. The method of mathematical induction allows us to confirm that our definitions of the medians and the center of gravity of an n-gon are meaningful for arbitrary n.

Problem 24. In the n-gon $A_1A_2 \ldots A_n$ let O_1 be the center of gravity of the $(n-1)$-gon $A_2A_3 \ldots A_n$, let O_2 be the center of gravity of the $(n-1)$-gon $A_1A_3 \ldots A_n$, etc., let O_n be the center of gravity of the $(n-1)$-gon $A_1A_2 \ldots A_{n-1}$. Prove that the n-gon $O_1O_2 \ldots O_n$ is similar to the given n-gon $A_1A_2 \ldots A_n$.

Hint. By the method of Example 26, O_1O_2 is parallel to A_1A_2 and

$$\frac{O_1O_2}{A_1A_2} = \frac{1}{n-1}.$$

Analogously, O_2O_3 is parallel to A_2A_3 and $\dfrac{O_2O_3}{A_2A_3} = \dfrac{1}{n-1}$, etc.

A line segment joining the center of gravity of a k-gon formed from any k of the vertices of an n-gon with the center of gravity of the $(n-k)$-gon formed from the remaining $n-k$ vertices is called a *median of the kth order of an n-gon* $(k < n)$. Thus, a median of the kth order is at the same time a median of the $(n-k)$th order. The medians of an n-gon defined in Example 26 might be called *medians of the first order*.

Problem 25. Prove that all of the medians of the kth order of an n-gon intersect at a common point, which divides each of them in the ratio $(n - k):k$.

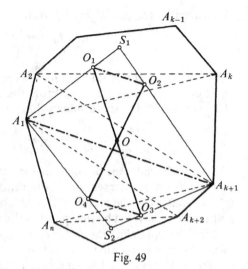

Fig. 49

Hint. Let S_1 and S_2 be the centers of gravity of the $(k - 1)$-gon $A_2A_3 \ldots A_k$ and the $(n - k - 1)$-gon $A_{k+2}A_{k+3} \ldots A_n$; let O_1 and O_2 be the centers of gravity of the k-gons $A_1A_2 \ldots A_k$ and $A_2A_3 \ldots A_{k+1}$; and let O_3 and O_4 be the centers of gravity of the $(n - k)$-gons $A_{k+1} \ldots A_n$ and $A_{k+2} \ldots A_nA_1$ (Fig. 49). Then

$$\frac{O_1S_1}{O_1A_1} = \frac{O_2S_1}{O_2A_{k+1}} = \frac{1}{k - 1},$$

and O_1O_2 is parallel to A_1A_{k+1};

$$\frac{O_3S_2}{O_3A_{k+1}} = \frac{O_4S_2}{O_4A_1} = \frac{1}{n - k - 1},$$

and O_3O_4 is parallel to A_1A_{k+1}. If now O is the point of intersection of the medians of the kth order O_2O_4 and O_1O_3, then from the similarity of the triangles OO_1O_2 and OO_3O_4 we have

$$\frac{OO_1}{OO_3} = \frac{OO_2}{OO_4} = \frac{O_1O_2}{O_3O_4} = \frac{\dfrac{1}{k}A_1A_{k+1}}{\dfrac{1}{n - k}A_1A_{k+1}} = \frac{n - k}{k}.$$

It is possible to prove that for arbitrary k the point of intersection of the medians of kth order of an n-gon coincides with its center of gravity.

Problem 26. Formulate the assertion of Problem 25 for $n = 4$, $k = 2$.

Answer. In an arbitrary quadrilateral the line segments joining the mid-points of pairs of opposite sides and the mid-points of the diagonals intersect in a common point, which bisects each of them.

19. EULER CIRCLES

The circle which passes through the mid-points of the three sides of a triangle (Fig. 50) is called the *Euler circle* of this triangle;

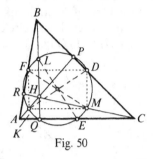

Fig. 50

it has a number of interesting properties. (For example, besides passing through the mid-points D, E, F of the sides, the Euler circle of the triangle ABC also passes through the feet P, Q, R of the altitudes AP, BQ, and CR and through the three points K, L, M which bisect the segments AH, BH, and CH of the altitudes between H, their intersection point, and the vertices;[1] for this reason the Euler circle is also often called the *nine-point circle* of the triangle.) Since the Euler circle of the triangle ABC circumscribes the triangle DEF, which is similar to ABC with similarity coefficient $\frac{1}{2}$, the radius of the Euler circle must be $R/2$, where R is the radius of the circumscribed circle of the initial triangle ABC. The concept of the Euler circle may be extended as follows to an arbitrary polygon inscribed in a circle.

---

[1] In the triangles ABH and CBH with the common base BH, the lines KF and DM join the mid-points of the sides, so that each is parallel to the line BH, and hence KF is parallel to DM. Similarly, in the triangles ABC and AHC with the common base AC, the lines FD and KM join the mid-points of the sides so that each is parallel to the line AC, and hence FD is parallel to KM. Since BQ is perpendicular to AC, the quadrilateral $KFDM$ (Fig. 50) is a rectangle; hence the line segments FM and DK are equal and bisect each other. In the same way it can be proved that the segment EL is equal to these segments also, and the mid-point of EL coincides with the common mid-point of FM and KD. From this it follows that the Euler circle, which passes through the points D, E, and F, also passes through the points K, L, and M (the center of this circle coincides with the common mid-point of the lines DK, EL, and FM, and its diameter is equal to the length of each of these equal segments. Further, since we have shown that K and D are end points of a diameter of the Euler circle and since $\angle KPD = 90°$, then the Euler circle passes through the point P; it can be shown similarly that it passes through the points Q and R, also.

Problem 27. 1. The *Euler circle of a chord A_1A_2* of a circle S with radius R is the circle of radius $R/2$ whose center is the mid-point of the chord A_1A_2 (Fig. 51a). The three Euler circles of the sides of a triangle $A_1A_2A_3$ inscribed in the circle S intersect at a common point O, which is the center of a circle of radius $R/2$ passing through the centers of the three Euler circles; this circle is the *Euler circle of the triangle $A_1A_2A_3$* (Fig. 51b) as defined above.

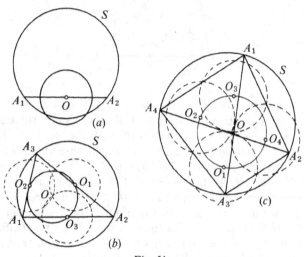

Fig. 51

2. Let us assume that we have defined the Euler circle for an *n*-gon inscribed in a circle S of radius R, and that its radius is $R/2$. Let us now consider an $(n + 1)$-gon $A_1A_2 \ldots A_{n+1}$ inscribed in a circle S. In that case the $n + 1$ Euler circles of the *n*-gons $A_2A_3 \ldots A_{n+1}, A_1A_3 \ldots A_{n+1}, \ldots, A_1A_2 \ldots A_n$ intersect at a common point, which is the center of a circle of radius $R/2$ which passes through the center of all of the $n + 1$ Euler circles; this circle is called the *Euler circle of the $(n + 1)$-gon $A_1A_2 \ldots A_{n+1}$* (see Fig. 51c, where the Euler circle for a quadrilateral is shown).

Hint. Let $A_1A_2A_3A_4$ be an arbitrary quadrilateral inscribed in the circle S. Because, for example, the Euler circle of the triangle $A_1A_2A_3$ passes through the three mid-points of the line segments H_4A_1, H_4A_2, H_4A_3, where H_4 is the intersection point of the altitudes of the triangle $A_1A_2A_3$ (see above), it follows that this Euler circle is centrally similar (homothetic) to the circle S with center of similitude at the point H_4 and coefficient of similitude $1/2$; therefore the mid-point of the segment H_4A_4 lies on this circle. Now all that remains is to note that the mid-points of the segments H_1A_1, H_2A_2, H_3A_3, and H_4A_4 coincide (where H_1, H_2, and H_3 are the intersection points of the altitudes of the corresponding triangles). This follows from the fact that, for example, the quadrilateral $A_1H_2H_1A_2$ is a parallelogram because A_1H_2 is parallel to A_2H_1 and perpendicular to A_3A_4 and $A_1H_2 =$ twice the distance from the center of S to A_3A_4.

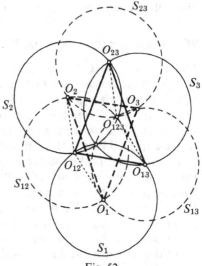

Fig. 52

Now let us assume the existence of an Euler circle for all k-gons for which the number of sides k is not greater than $n \geq 4$, and let us consider an $(n + 1)$-gon $A_1A_2 \ldots A_nA_{n+1}$ inscribed in a circle S. In order to prove that the Euler circles $S_1, S_2, \ldots, S_{n+1}$ for the n-gons $A_2A_3 \ldots A_{n+1}$, $A_1A_3A_4 \ldots A_{n+1}, \ldots, A_1A_2 \ldots A_n$ intersect at a common point, it is sufficient to prove that any three of them intersect at a common point, for example, S_1, S_2, and S_3.[1] Let S_{12}, S_{13}, and S_{23} denote the Euler circles for the $(n-1)$-gons $A_3A_4 \ldots A_{n+1}$, $A_2A_4A_5 \ldots A_{n+1}$, and $A_1A_4A_5 \ldots A_{n+1}$, and let O_{12}, O_{13}, and O_{23} be their centers; further, let O_1, O_2, and O_3 be the centers of the circles S_1, S_2, and S_3, and let O_{123} be the center of the Euler circle S_{123} for the $(n-2)$-gon $A_4A_5 \ldots A_{n+1}$. In this case we arrive at Fig. 52, from which it is not difficult to see that $\triangle\, O_1O_2O_3 \cong \triangle\, O_{23}O_{13}O_{12}$. (To prove the equality of the sides O_1O_2 and $O_{23}O_{13}$ of these triangles, it suffices to consider the triangles $O_1O_2O_{12}$ and $O_{23}O_{13}O_{123}$ in which

$$O_{12}O_1 = O_{12}O_2 = O_{123}O_{23} = O_{123}O_{13} = \frac{R}{2},$$

$$\angle\, O_1O_{12}O_2 = \angle\, O_1O_{12}O_{123} + \angle\, O_{123}O_{12}O_2$$
$$= 2\, \angle\, O_{13}O_{12}O_{123} + 2\, \angle\, O_{123}O_{12}O_{23}$$
$$= 2\, \angle\, O_{13}O_{12}O_{23},$$

and
$$\angle\, O_{23}O_{123}O_{13} = 2\, \angle\, O_{13}O_{12}O_{23},$$

since the inscribed and central angles for the circle circumscribed about $O_{12}O_{13}O_{23}$ subtend the same arc; in the same way it can be proved that $O_1O_3 = O_{23}O_{12}$ and $O_2O_3 = O_{13}O_{12}$.) From the fact that $\triangle\, O_1O_2O_3 \cong \triangle\, O_{23}O_{13}O_{12}$ and the circles S_{23}, S_{13}, and S_{12} intersect at a common point O_{123}, it follows that the circles S_1, S_2, and S_3 also intersect at a common point.

[1] For if every three of $n \geq 5$ circles (no two of which coincide) intersect at a common point, then all of the circles intersect at the same point. (For $n = 4$, this is not true.)

Problem 28. Let $A_1 A_2 \ldots . A_n$ be an arbitrary n-gon inscribed in a circle S. Prove that the center of gravity of the n-gon (see Example 26) lies on the segment joining the center of S with the center of the Euler circle of the n-gon and divides it in the ratio $(n-2):2$.[1]

20. MEAN CIRCLES (OPTIONAL)

EXAMPLE 27.   1. Let l_1, l_2, l_3, and l_4 be four straight lines in general position, that is, such that no two of them are parallel and no three of them pass through a common point. Let O_1 be the center of the circle circumscribed about the triangle formed by the straight lines l_2, l_3, and l_4; let O_2 be the center of the circle circumscribed about the triangle formed by the straight lines l_1, l_3, and l_4, etc. Then the four points O_1, O_2, O_3, and O_4 lie on a circle, which is called the *mean circle* of the four straight lines l_1, l_2, l_3, and l_4 (Fig. 53).

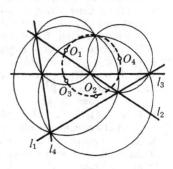

Fig. 53

2. Suppose that the mean circle of n straight lines has been defined, and that we are given $n+1$ straight lines l_1, l_2, l_3, $\ldots$, l_{n+1} in general position. Let O_1 denote the center of the mean circle of the n straight lines l_2, l_3, $\ldots$, l_{n+1}, O_2 the center of the mean circle of the n straight lines l_1, l_3, $\ldots$, l_{n+1}, etc. Then the $n+1$ points O_1, O_2, O_3, $\ldots$, O_{n+1} lie on a circle, which is called the *mean circle* of the $n+1$ straight lines l_1, l_2, l_3, $\ldots$, l_{n+1}.

Proof. 1. Let l_1, l_2, l_3, and l_4 be four straight lines in general position (Fig. 54), A_{12} be the point of intersection of l_3 and l_4, A_{13} the point of intersection of l_2 and l_4, etc. Let O_1 be the center of the circle C_1 circumscribed about the triangle formed by the straight lines l_2, l_3, and l_4, etc. First of all, let us prove that the circles C_1, C_2, C_3, and C_4 intersect at a common point M. Indeed, if M is the point of intersection (distinct from A_{12}) of C_1 and C_2, then

$$\angle A_{13} M A_{12} = \angle A_{13} A_{14} A_{12} = \angle \text{ between } l_2 \text{ and } l_3;$$

$$\angle A_{12} M A_{23} = \angle A_{12} A_{24} A_{23} = \angle \text{ between } l_3 \text{ and } l_1.$$

[1] The solution of this problem may be found in Part II of the book by I. M. Yaglom mentioned on p. 49 (see the solution of Problem 52c).

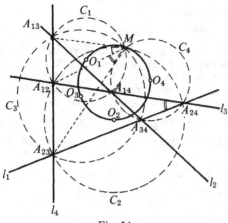

Fig. 54

From this it follows that $\angle A_{13}MA_{23} = \angle$ between l_2 and l_1 $= \angle A_{13}A_{34}A_{23}$; that is, the circle C_3 passes through M. In the same way we can prove that C_4 also passes through M.

Now we are ready to prove that the points O_1, O_2, O_3, and O_4 lie on a circle. Let us consider the three circles C_1, C_2, and C_3, which pass through the same point M; C_1 and C_3 intersect again at the point A_{13}, and C_2 and C_3 intersect again at the point A_{23}. Then we have[1]

$$\angle O_1O_3O_2 = \angle A_{13}MA_{23} = \angle A_{13}A_{34}A_{23} = \angle \text{ between } l_2 \text{ and } l_1.$$

In exactly the same way we prove that $\angle O_1O_4O_2 = \angle$ between l_2 and $l_1 = \angle O_1O_3O_2$,[1] from which the desired assertion follows.

2. Let us assume that our assertion has been proved for n straight lines; we may also consider as proved that the arc of the mean circle for the n straight lines $l_1, l_2, \ldots, l_n$ between the centers O_1 and O_2 of the mean circles of the $n-1$ straight lines $l_2, l_3, \ldots, l_n$ and the $n-1$ straight lines $l_1, l_3, \ldots, l_n$ is twice the angle between the straight lines l_1 and l_2 (see the last part of part 1). Let us now consider the $n+1$ straight lines $l_1, l_2, \ldots, l_{n+1}$ in general position; let O_1 be the center of the circle C_1 which is the mean circle for the n straight lines $l_2, l_3, \ldots, l_{n+1}$, etc., O_{12} the center of the circle C_{12} which is the mean circle for the $n-1$ straight lines $l_3, l_4, \ldots, l_{n+1}$, etc. Let us prove that the circles $C_1, C_2, \ldots, C_{n+1}$ intersect at a common point M. Indeed, let M be the point of intersection different from O_{12} of the circles C_1 and C_2. Then we have[1]

$$\angle O_{13}MO_{12} = \tfrac{1}{2} \text{ arc } O_{13}O_{12} = \angle \text{ between } l_2 \text{ and } l_3;$$

$$\angle O_{12}MO_{23} = \tfrac{1}{2} \text{ arc } O_{12}O_{23} = \angle \text{ between } l_3 \text{ and } l_1.$$

[1] More exactly, these angles are equal or supplementary.

From this it follows that[1]

$$\angle\, O_{13}MO_{23} = \angle \text{ between } l_2 \text{ and } l_1 = \angle\, O_{13}O_{34}O_{23},$$

that is, the circle C_3 passes through M. In exactly the same way we can prove that each of the remaining circles $C_4, C_5, \ldots, C_{n+1}$ also passes through M.

Let us consider next the three circles C_1, C_2, and C_3 which pass through the point M; O_{13} is the second point of intersection of C_1 and C_3, O_{23} the second point of intersection of C_2 and C_3. Then we have

$$\angle\, O_1O_3O_2 = \angle\, O_{13}MO_{23} = \angle\, O_{13}O_{34}O_{23} = \angle \text{ between } l_2 \text{ and } l_1.[1]$$

In exactly the same way we can show that for an arbitrary point O_i ($i = 4, 5, \ldots, n + 1$)

$$\angle\, O_{13}MO_{23} = \angle \text{ between } l_2 \text{ and } l_1 = \angle\, O_{13}O_{34}O_{23};$$

from which it follows that all of the points $O_1, O_2, O_3, O_4, \ldots, O_{n+1}$ lie on a circle.

In the formulation of Example 27 we could replace "circumscribed circles" everywhere by "inscribed circles." But then a complication would arise due to the fact that while the circumscribed circle of a triangle (a circle passing through all of its vertices) is uniquely determined, the inscribed circle (the circle tangent to all of the sides of the triangle) may be any one of four circles (each of the three sides is tangent to one inscribed and three escribed circles). In order to avoid this difficulty we can proceed in the following way. Let us introduce *directed* straight lines and circles, using arrows to show the direction of motion along each. Further, let us regard a directed straight line and a directed circle as tangent only if their directions at the point of tangency coincide. Then there will be a *unique* directed circle tangent to three given directed straight lines l_1, l_2, and l_3 which do not intersect at a common

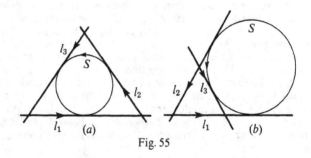

Fig. 55

point (Fig. 55a,b), and this circle will be the directed inscribed circle of the triangle formed by l_1, l_2, and l_3.

[1] These angles may be equal or supplementary.

Problem 29. The definition of the orthocenter of a polygon inscribed in a circle.

1. The orthocenter of a triangle is defined as the point of intersection of its altitudes.

2. Suppose we have defined the orthocenter of an n-gon $A_1A_2 \ldots A_n$ inscribed in a circle S, and suppose we are given an $(n + 1)$-gon $A_1A_2 \ldots A_nA_{n+1}$ inscribed in S. Let the orthocenters of the $n + 1$ polygons $A_2A_3 \ldots A_{n+1}$, $A_1A_3 \ldots A_{n+1}, \ldots, A_1A_2, \ldots, A_n$ be denoted by $H_1, H_2, \ldots, H_{n+1}$, respectively. Then prove that circles having the same radius as S and having their centers at points $H_1, H_2, \ldots, H_{n+1}$, intersect at one point H. This point is called the orthocenter of the $(n + 1)$-gon $A_1A_2 \ldots A_{n+1}$. (Thus, the orthocenter of the quadrilateral $A_1A_2A_3A_4$ is shown in Fig. 56.)

We leave the solution of Problem 29 to the reader.

The orthocenters of polygons inscribed in a circle have certain properties that are similar to properties of the orthocenters of triangles. We shall not discuss these properties, even though they must be proved by induction (since the orthocenter of a polygon is defined by induction).

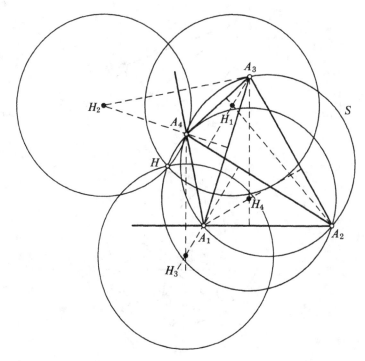

Fig. 56

21. CENTRAL POINTS AND CENTRAL CIRCLES

Problem 30. 1. The point of intersection of two (intersecting) straight lines is called the *central point of the two straight lines* (Fig. 57a).

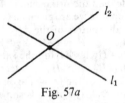

Fig. 57a

The *central circle* of three straight lines in general position (see Example 27) is the circle which passes through the central point of every pair of these straight lines (Fig. 57b).

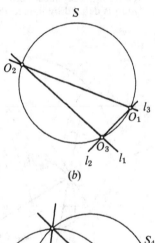

(b)

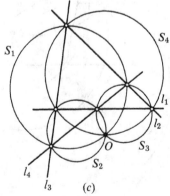

(c)

Fig. 57

Suppose now that we are given four straight lines l_1, l_2, l_3, and l_4 in general position. Let S_1 denote the central circle of the three straight lines l_2, l_3, l_4; S_2 the central circle of the three straight lines l_1, l_3, l_4; etc. Then the four circles S_1, S_2, S_3, and S_4 intersect at a common point O, which is called the *central point of the four straight lines* l_1, l_2, l_3, l_4 (Fig. 57c).

2. Let us assume that we have already determined the central circle of $2n - 1$ straight lines and the central point of $2n$ straight lines, and suppose that we are given $2n + 1$ straight lines l_1, l_2, ..., l_{2n}, l_{2n+1} in general position. Let A_1 denote the central point of the $2n$ straight lines l_2, l_3, ..., l_{2n}, l_{2n+1}; A_2 the central point of the $2n$ straight lines l_1, l_3, ..., l_{2n}, l_{2n+1}; etc.; A_{2n+1} the central point of the $2n$ straight lines l_1, l_2, ..., l_{2n}. Then the points A_1, A_2, ..., A_{2n+1} lie on a circle, which we shall call the *central circle of the $2n + 1$ straight lines* l_1, l_2, ..., l_{2n}, l_{2n+1}.

Finally, suppose that we are given $2n + 2$ straight lines l_1, l_2, ..., l_{2n+1}, l_{2n+2} in general position. Let S_1 denote the central circle of the $2n + 1$ straight lines l_2, l_3, ..., l_{2n+1}, l_{2n+2}; S_2 the central circle of the $2n + 1$ straight lines l_1, l_3, ..., l_{2n+1}, l_{2n+2}; etc., S_{2n+2} the central circle of the $2n + 1$ straight lines l_1, l_2, ..., l_{2n+1}.

Then the circles S_1, S_2, ..., S_{2n+1}, S_{2n+2} intersect at a common point, which we shall call the *central point of the $2n + 2$ straight lines* l_1, l_2, ..., l_{2n+1}, l_{2n+2}.

Problem 31. 1. Suppose we are given three straight lines l_1, l_2, and l_3 in general position. The center of the circumscribed circle of the triangle formed by these lines is called the *central point of the three straight lines*.

Let us now consider four straight lines l_1, l_2, l_3, and l_4 in general position. Let A_1 denote the central point of the three straight lines l_2, l_3, l_4; A_2 the central point of the three straight lines l_1, l_3, l_4; etc. Then the four points A_1, A_2, A_3, and A_4 lie on a circle (see Example 27, above), which is called the *central circle of the four straight lines* l_1, l_2, l_3, l_4.

2. Let us assume that we have already determined the central point for $2n - 1$ straight lines and the central circle for $2n$ straight lines, and suppose that we are given $2n + 1$ straight lines l_1, l_2, ..., l_{2n}, l_{2n+1} in general position. Let S_1 denote the central circle of the $2n$ straight lines l_2, l_3, ..., l_{2n}, l_{2n+1}; S_2 the central circle of the $2n$ straight lines l_1, l_3, ..., l_{2n}, l_{2n+1}; etc.; S_{2n+1} the central circle of the $2n$ straight lines l_1, l_2, ..., l_{2n}. Then the circles S_1, S_2, ..., S_{2n+1} intersect in a common point, which we shall call the *central point of the $2n + 1$ straight lines* l_1, l_2, ..., l_{2n}, l_{2n+1}.

Finally, suppose that we are given $2n + 2$ straight lines l_1, l_2, ..., l_{2n+2} in general position. Let A_1 denote the central point of the $2n + 1$ straight lines l_2, l_3, ..., l_{2n+2}; A_2 the central point of the $2n + 1$ straight lines l_1, l_3, ..., l_{2n+2}; etc.; A_{2n+2} the central point of the $2n + 1$ straight lines l_1, l_2, ..., l_{2n+1}.

Then the points A_1, A_2, ..., A_{2n+2} lie on a circle, which we shall call the *central circle of the $2n + 2$ straight lines* l_1, l_2, ..., l_{2n+2}.

Hint. The proof of the propositions formulated here is completely analogous to the proof of the propositions which comprise Problem 30.

22. LINEAR ELEMENTS

By a *linear element* we mean a point A, together with straight line a passing through the point A in a given direction. A linear element will be denoted by the symbol (A, a). We say that the n linear elements (A_1, a_1), (A_2, a_2), ..., (A_n, a_n) are *concyclic* if the straight lines $a_1, a_2, ..., a_n$ are in general position (see Example 27, above) and if the n points $A_1, A_2, ..., A_n$ lie on a circle.

Problem 32. 1. If A_1 and A_2 are distinct points and if the straight lines a_1 and a_2 intersect, then the circle which passes through the points A_1, A_2 and the point of intersection of a_1 and a_2 is called the *directing circle of the two linear elements* (A_1, a_1) *and* (A_2, a_2); see Fig. 58a. Given three pairs of linear elements (A_1, a_1) and (A_2, a_2), (A_1, a_1) and (A_3, a_3), and (A_2, a_2) and (A_3, a_3) with the straight lines a_1, a_2, a_3 in general position and with the points A_1, A_2, A_3 distinct, the directing circles of these elements will intersect in a common point, called the *directing point of the three linear elements* (A_1, a_1), (A_2, a_2), *and* (A_3, a_3); see Fig. 58b.

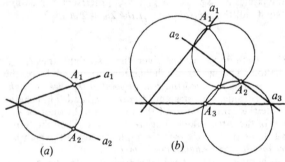

Fig. 58

2. Suppose that we have already determined the directing circle for $2n - 2$ concyclic linear elements and the directing point for $2n - 1$ concyclic linear elements. Let us consider $2n$ concyclic linear elements. In this case the $2n$ directing points for every collection of $2n - 1$ of the linear elements will lie on a circle, which is called the *directing circle of the $2n$ concyclic linear elements*. Further, if we consider $2n + 1$ concyclic linear elements, then the collections of $2n$ of the linear elements will define $2n + 1$ directing circles which all intersect at a common point, which is called the *directing point of the $2n + 1$ concyclic linear elements*.

6. Induction on the Number of Dimensions

23. SPACES OF VARIOUS DIMENSIONS

In a course in solid geometry we are struck by a certain analogy between the theorems of solid geometry and those of plane geometry. Thus, the properties of a parallelepiped are similar to the properties of a parallelogram in many respects. (Compare, for example, the theorems, "The opposite faces of a parallelepiped are congruent and the diagonals intersect in a common point which bisects each of them," and "The opposite sides of a parallelogram are equal and the diagonals intersect in a point which bisects each of them.") The properties of a sphere are similar to those of a circle. (Compare, for example, the theorems, "A plane tangent to a sphere is perpendicular to the radius at the point of tangency," and "A line tangent to a circle is perpendicular to the radius at the point of tangency.") But at the same time, there are also essential differences between the properties of plane figures and space figures. The principal difference here consists of the fact that figures in the plane have two dimensions ("length" and "width") while those in space have three dimensions ("length," "width," and "height"). Accordingly, the position of a point in the plane is completely determined by two coordinates x and y (Fig. 59a), while in order to

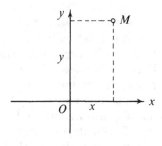

Fig. 59a

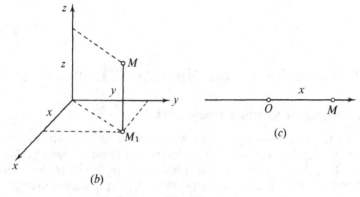

Fig. 59

determine the position of a point in space it is necessary to specify three coordinates x, y, and z (Fig. 59b). It is for this reason that our ordinary space is frequently called *three-dimensional space* (or "space of three dimensions"), while one speaks of the plane as *two-dimensional space* (or "space of two dimensions"). This terminology may be extended still further. The position of a point on a straight line is completely determined by a single coordinate x (Fig. 59c); this is because on a straight line all figures (segments) have only one dimension ("length"). For this reason a straight line is called *one-dimensional space;* thus, the number of dimensions of a space may be one, two, or three.

Theorems in solid geometry are usually more complex than the corresponding propositions in plane geometry; in their turn, the properties of plane figures are, of course, far more complex than the properties of figures on a straight line (segments). At the same time, proofs of "three-dimensional" theorems (that is, theorems in solid geometry) often rest substantially on knowledge of the corresponding "two-dimensional" propositions (that is, propositions in plane geometry). Thus, for example, the proof that the diagonals of a parallelepiped are bisected by their common intersection point makes use of the corresponding property of the diagonals of a parallelogram. Proofs of "two-dimensional" theorems, in their turn, are sometimes based on analogous proofs of one-dimensional theorems. This circumstance makes it possible, in certain geometric problems, to use *induction on the number of dimensions*, which consists in passing successively from one-dimensional to two-dimensional, and then to three-dimensional space. Examples of this are given in this chapter. Induction on the number of dimensions is often used simultaneously with ordinary induction, and sometimes ordinary induction may be substituted for it (section 26).

In studying the examples and solving the problems of this chapter, it must be kept in mind that a circle in the plane (that is, the locus of the points equidistant from a given point O, Fig. 60b) corresponds to a sphere in space (the surface of a ball, Fig. 60c), while it corresponds to a pair of points on a straight line equidistant from a given point O (Fig. 60a); a circular disc in the plane corresponds to a solid sphere in space, and a line segment on a straight line.

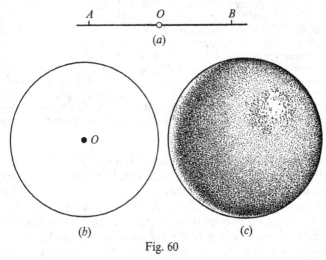

Fig. 60

Finally, a triangle ABC in the plane (Fig. 61b) corresponds to a tetrahedron in space (that is, an arbitrary triangular pyramid having four vertices A, B, C, and D, Fig. 61c), and on the line to a line segment AB having two "vertices" A and B (Fig. 61a).

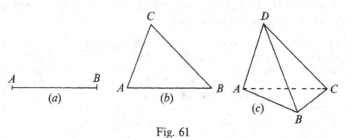

Fig. 61

We should note that the question of what proposition in solid geometry should correspond to a given theorem in plane geometry does not have a unique answer, generally speaking. Sometimes it is convenient to regard a triangle in a plane as corresponding not to a tetrahedron (a figure having one more dimension), but rather to the same triangle only situated in space. Analogously, we may regard a straight line in the plane as corresponding in space to either a straight line or a plane. Thus, we may obtain different "three-dimensional analogues" for one and the same two-dimensional theorem. For example, the theorem "The sum of the squares of the distances of a point M in the plane from all of the vertices of a regular n-gon with center O inscribed in a circle of radius R is equal to $n(R^2 + OM^2)$" corresponds to the following two theorems in solid geometry: "The sum of the squares of the distances of a point M in space from all the vertices of a regular n-gon with center O inscribed in a circle of radius R is equal to $n(R^2 + OM^2)$," and "The sum of the squares of the distances of a point M in space from all the vertices of a regular polyhedron with n vertices inscribed in a sphere with center O and radius R is equal to $n(R^2 + OM^2)$"; both of these theorems are true and both are deduced from the corresponding "two-dimensional" theorem, so that in deducing each, induction on the number of dimensions is applied. We shall not dwell on this question in great detail, but leave it to the reader to compare for himself the transition from "one-dimensional" to "two-dimensional" and "three-dimensional" theorems, for example, in Examples 29 and 37 below, on the one hand, and Examples 35 and 36, on the other hand.

In this chapter we shall consider computation using induction on the number of dimensions, proofs using induction on the number of dimensions, the determination of geometric loci by using induction on the number of dimensions, and definitions using induction on the number of dimensions. We shall not consider using induction on the number of dimensions to solve problems in construction, because the very question of constructions in space is not sufficiently well defined. In all cases the "three-dimensional" proposition (in solid geometry) is considered basic, although frequently the corresponding result in the "two-dimensional" (plane) case is more interesting; in these cases the transition from the two-dimensional to the three-dimensional case will only be indicated and not carried out in full detail.

In modern mathematics and physics an important part is played by the concept of *n-dimensional space,* wherein the position of a point is determined by n coordinates; here n is an arbitrary positive integer; in particular, n may be greater than three. Properties of figures in n-dimensional space are often proved by means of mathematical induction on the number of dimensions of the space; in particular, this method enables one to extend all the results of this section to n-dimensional space. However, this is beyond the scope of the present booklet.

24. COMPUTATION USING INDUCTION ON THE NUMBER OF DIMENSIONS

EXAMPLE 28. Into how many parts is space divided by n planes "in general position" (that is, such that every three of the planes intersect in a common point, but no four of them have a point in common)?

Let us examine the following sequence of problems:

A. Determine into how many parts a straight line is divided by n points.

SOLUTION. Let $F_1(n)$ denote the number of parts; obviously $F_1(n) = n + 1$.

B. Into how many parts is a plane divided by n straight lines "in general position" (each pair of which intersect but no three of which have a common point)?

SOLUTION. 1. One straight line divides the plane into two parts.

2. Let us assume that we know $F_2(n)$, the number of parts into which the plane is divided by n straight lines in general position, and consider $n + 1$ straight lines in general position. The first n of these lines divide the plane into $F_2(n)$ parts; the $(n + 1)$st straight line l, by the conditions, intersects the remaining n straight lines in n distinct points; these points divide the straight line l into $F_1(n) = n + 1$ parts (see part A). Consequently, the straight line l intersects $n + 1$ of the parts of the plane previously obtained, and hence $F_2(n)$ is increased by $F_1(n)$ new parts. Thus,

$$F_2(n + 1) = F_2(n) + F_1(n) = F_2(n) + (n + 1). \qquad (13)$$

Replacing n by $n - 1, n - 2, \ldots, 2, 1$ in equality (13), we obtain

$$F_2(n) = F_2(n - 1) + n,$$
$$F_2(n - 1) = F_2(n - 2) + (n - 1),$$
$$\cdots$$
$$F_2(3) = F_2(2) + 3,$$
$$F_2(2) = F_2(1) + 2.$$

Let us add all these equalities; since $F_2(1) = 2$, we have

$$F_2(n) = F_2(1) + [n + (n - 1) + \cdots + 2]$$
$$= 1 + [n + (n - 1) + \cdots + 2 + 1],$$

and, finally

$$F_2(n) = 1 + \frac{n(n + 1)}{2} = \frac{n^2 + n + 2}{2}$$

(see formula (2) of the Introduction).

C. The problem formulated at the beginning of this example.

SOLUTION. 1. One plane divides space into two parts.

2. Let us assume that we know $F_3(n)$, the number of parts into which space is divided by n planes in general position, and consider $n + 1$ planes in general position. The first n of these divide space into $F_3(n)$ parts; the $(n + 1)$st plane π intersects these n planes in n straight lines in general position and, consequently, is divided by them into $F_2(n) = (n^2 + n + 2)/2$ parts (see part B). Thus, we obtain the relationship

$$F_3(n + 1) = F_3(n) + F_2(n) = F_3(n) + \frac{n^2 + n + 2}{2}. \quad (14)$$

Replacing n by $n - 1, n - 2, \ldots, 2, 1$ in equality (14), we obtain

$$F_3(n) = F_3(n - 1) + \frac{(n - 1)^2 + (n - 1) + 2}{2},$$

$$F_3(n - 1) = F_3(n - 2) + \frac{(n - 2)^2 + (n - 2) + 2}{2},$$

$$\cdots$$

$$F_3(3) = F_3(2) + \frac{2^2 + 2 + 2}{2},$$

$$F_3(2) = F_3(1) + \frac{1^2 + 1 + 2}{2}.$$

Adding all of these equalities, we obtain

$$F_3(n) = F_3(1) + \frac{1}{2}[(n - 1)^2 + (n - 2)^2 + \cdots + 1^2]$$

$$+ \frac{1}{2}[(n - 1) + (n - 2) + \cdots + 1] + \frac{1}{2}\underbrace{[2 + 2 + \cdots + 2]}_{n - 1 \text{ terms}},$$

or, finally, taking into account formulas (2) and (3) of the Introduction and the fact that $F_3(1) = 2$,

$$F_3(n) = 2 + \frac{n(n - 1)(2n - 1)}{12} + \frac{(n - 1)n}{4} + (n - 1)$$

$$= \frac{(n + 1)(n^2 - n + 6)}{6}.$$

Problem 33. Into how many parts is space divided by n spheres every pair of which intersect?

Hint. Let us consider the following sequence of problems:

A. Into how many parts is a straight line divided by n "one-dimensional circles," that is, n pairs of points (see section 23)?
Answer. $2n$ points divide the straight line into $2n + 1$ parts.

A′. Find $\Phi_1(n)$, the number of parts into which a circle is divided by n pairs of points lying on this circle.
Answer. $\Phi_1(n) = 2n$.

B. Find $\Phi_2(n)$, the number of parts into which a plane is divided by n pairwise intersecting circles.
SOLUTION. Since n of the circles intersect the $(n + 1)$st circle in n pairs of points and, consequently, divide it into $\Phi_1(n) = 2n$ parts (see part A′), then the $(n + 1)$st circle intersects $\Phi_1(n) = 2n$ of the $\Phi_2(n)$ parts into which the n circles divide the plane. From this we obtain the equality

$$\Phi_2(n + 1) = \Phi_2(n) + \Phi_1(n)$$

$$= \Phi_2(n) + 2n.$$

Using this equality and the fact that $\Phi_2(1) = 2$, we have

$$\Phi_2(n) = n^2 - n + 2.$$

B′. Into how many parts is the surface of a sphere divided by n pairwise intersecting circles situated on this sphere?
Answer. Into $\Phi_2(n) = n^2 - n + 2$ parts.

C. The initial problem.
SOLUTION. Since n spheres intersect the $(n + 1)$st sphere in n circles and, consequently, divide its surface into $\Phi_2(n) = n^2 - n + 2$ parts (see part B′), then if the n pairwise intersecting spheres divide space into $\Phi_3(n)$ parts, $n + 1$ spheres will divide space into

$$\Phi_3(n + 1) = \Phi_3(n) + \Phi_2(n)$$

$$= \Phi_3(n) + (n^2 - n + 2)$$

parts. From this and the fact that $\Phi_3(1) = 2$, we obtain

$$\Phi_3(n) = \frac{n(n^2 - 3n + 8)}{3}.$$

25. PROOFS USING INDUCTION ON THE NUMBER OF DIMENSIONS

EXAMPLE 29.   A tetrahedron with vertices numbered 1, 2, 3, and 4 is divided into smaller tetrahedrons in such a way that each pair of tetrahedrons which result from the subdivision either has no points in common, or has a vertex in common, or has an edge in common (but not a part of an edge), or has a face in common (but not a part of a face). All of the vertices of the small tetrahedrons obtained in this way are numbered using the digits 1, 2, 3, and 4. Moreover, all the vertices lying on a face of the large tetrahedron are numbered using only the three digits which number the vertices of that face, and all of the vertices lying on any edge of the large tetrahedron are numbered using only the two digits which number the ends of this edge. Prove that there is at least one tetrahedron for which all four vertices bear different digits.

Let us consider the following sequence of problems:

A.   A segment with ends numbered 1 and 2 is divided into several nonintersecting smaller segments, and all of the points of the subdivision are numbered with the digits 1 or 2 (Fig. 62a). Prove that there is at least one segment of the subdivision for which the two ends bear different digits.

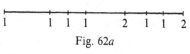

Fig. 62a

Proof. Let us prove that the number of segments numbered 1 2 is odd; from this it will follow that there exists at least one such segment (since zero is an even number). Let k denote the number of 1's in the numbering of the subdivision, and let A denote the number of segment-ends of the subdivision numbered 1. The number A is obviously odd because each 1 standing at an interior point of the original segment is an end point for two segments of the subdivision, while only a single 1 belongs to only one segment of the subdivision, namely the 1 at the end of the original segment; consequently,

$$A = 2k + 1.$$

On the other hand, if we let p denote the total number of segments in our subdivision numbered 1 1 and let q denote the number of segments numbered 1 2, then A, the number of the ends numbered 1, will be

$$A = 2p + q.$$

From the equality

$$2k + 1 = 2p + q$$

it follows that q is odd.

B. A triangle with vertices numbered 1, 2, and 3 is divided into smaller triangles in such a way that any two of the triangles which result from the subdivision either have no point in common, or have a common vertex, or have a common side (but not a part of a side). All vertices of the triangles of the subdivision are numbered with the digits 1, 2, and 3. Moreover, any vertex lying on a side of the larger triangle is numbered using one of the two digits which number the ends of that side (Fig. 62b). Prove that there is at least one triangle of the subdivision all of whose vertices are numbered with different digits (shaded in Fig. 62b).

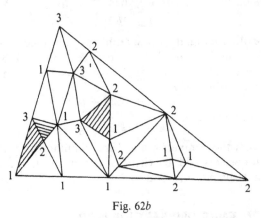

Fig. 62b

Proof. Let us show that the number of triangles numbered 1 2 3 is odd. To do this let us count A, the total number of sides of triangles of the subdivision numbered 1 2. Let k denote the number of interior segments numbered 1 2, and let l denote the number of such segments lying on the 1 2 side of the larger triangle (on the other two sides of the larger triangle there can be no 1 2 segments). Since each of the former k segments belongs to two triangles of the subdivision and each of the latter l segments belongs to one, then

$$A = 2k + l.$$

On the other hand, let p be the number of triangles of the subdivision with vertices numbered either 1 2 2 or 1 2 1, and let q be the number of triangles of the subdivision with vertices numbered 1 2 3. Then since each of the former p triangles has two 1 2 sides, and each of the latter q triangles has one such side, we have

$$A = 2p + q.$$

From the equality

$$2k + l = 2p + q$$

it follows that q and l are both even or else both odd. But the number l is odd by virtue of the proposition established in A, and, consequently, q is odd.

C. The proposition formulated at the beginning of this example.

Proof. Let A be the number of faces of tetrahedrons of the subdivision numbered with the digits 1, 2, 3. If k is the number of such faces lying inside the larger tetrahedron and l is the number of such faces on its 1 2 3 surface, then

$$A = 2k + l.$$

On the other hand, if p is the number of tetrahedrons of the subdivision bearing the digits 1 1 2 3 or 1 2 2 3 or 1 2 3 3, and q is the number of tetrahedrons of the subdivision bearing the digits 1 2 3 4, then, obviously,

$$A = 2p + q.$$

From the equality

$$2k + l = 2p + q$$

it follows that the numbers q and l are simultaneously even or odd. But the number l is odd, by virtue of the proposition established in B; consequently, q is also odd.

26. PROOFS USING ORDINARY INDUCTION

In section 23 we pointed out that induction on the number of dimensions may sometimes be replaced by ordinary induction. We shall give a number of appropriate examples.

EXAMPLE 30. Prove the proposition formulated in Example 29A by induction on n, the number of segments in the subdivision.

Proof. 1. For $n = 1$ the assertion is obvious.

2. Let us assume that our assertion has been proved for an arbitrary subdivision of the segment into n smaller segments, and suppose that we are given a subdivision of the segment 1 2 into $n + 1$ smaller segments. If not all of these segments are numbered 1 2, then we can find a segment both ends of which are numbered alike, for example, 1 1. If we shrink this segment to a point, we obtain a subdivision of the segment 1 2 into n smaller segments. By the inductive hypothesis this subdivision, and hence also the initial subdivision, has at least one segment numbered 1 2.

EXAMPLE 31. Prove the proposition formulated in Example 29B by induction on *n*, the number of triangles in the subdivision.

Proof. 1. For $n = 1$ the assertion is obvious; for $n = 2$ it is easily verified.

2. Let us assume that our assertion has been proved for an arbitrary subdivision of the triangle 1 2 3 into *n* or fewer triangles, and suppose that we have a subdivision of the triangle into $n + 1$ triangles. If not all of these triangles are numbered 1 2 3, then there will be some triangle for which two of the vertices have the same number, for example, 1. Either two triangles of the subdivision lie on this side 1 1 (if the side lies interior to the larger triangle, Fig. 63*a*) or only one triangle (if this side lies on a side of the larger triangle, Fig. 63*b*). If we shrink the segment 1 1 to a point, we obtain a new subdivision of the triangle 1 2 3 into $n - 1$ triangles (in the first case, Fig. 63*c*) or *n* triangles (in the second case, Fig. 63*d*). By the inductive hypothesis this subdivision (and hence also the initial subdivision) has at least one triangle with vertices numbered 1 2 3.

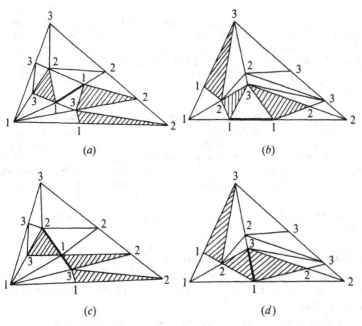

(a)

(b)

(c)

(d)

Fig 63

Problem 34. Prove the theorem of Example 29C by induction on *n*, the number of tetrahedrons in the subdivision.

Hint. The proof is analogous to the proof of the assertion in Example 31.

The proposition in Example 29 can be made even more exact. Let us introduce the concept of an "orientation" of the tetrahedron whose vertices are numbered 1 2 3 4, distinguishing tetrahedrons for which the circuit around the faces 1, 2, 3 from vertex 1, to vertex 2, and then to vertex 3 is clockwise viewed from vertex 4, from those for which this circuit viewed from vertex 4 is counterclockwise. Then the following proposition is true.

Problem 35. Prove that under the conditions of Example 29, the number of tetrahedrons in the subdivision numbered 1 2 3 4 and having the same orientation as the larger tetrahedron, will be greater by exactly one than the number of tetrahedrons 1 2 3 4 with the opposite orientation.

Hint. Consider the following sequence of problems:

A. Under the conditions of Example 29A let us distinguish between the segment 1 2 for which the direction from vertex 1 to vertex 2 coincides with that of the larger segment, and the segments 1 2 for which the direction from 1 to 2 is opposite to that of the larger segment. Prove that the number of segments of the first type is greater by 1 than the number of segments of the second type.

B. Let us say that the triangle 1, 2, 3 (see Example 29B) has clockwise orientation (or counterclockwise orientation) if the circuit around its vertices from vertex 1 to vertex 2 and then to vertex 3 is clockwise (or counterclockwise). Prove that the number of triangles of the subdivision numbered 1 2 3 with the same orientation as the larger triangle, is greater by exactly one than the number of the remaining triangles of the subdivision numbered 1 2 3.

C. The proposition formulated at the beginning of this problem.

EXAMPLE 32. Let *n* solid spheres be given in space such that every four of them have a common point. Prove that all of these spheres intersect, that is, there exists at least one point belonging to all of the spheres.

Let us consider the following sequence of problems:

A. On a straight line let n segments be given such that every pair intersect. Prove that all of the segments intersect, that is, there exists at least one point common to all of the segments.

Proof. 1. For $n = 2$ the assertion is obvious.

2. Let us assume that our assertion has been proved for n arbitrary segments, and suppose that on a straight line we are given $n + 1$ segments

$$l_1, l_2, \ldots, l_n, l_{n+1}$$

such that every pair intersect. By the inductive hypothesis the n segments

$$l_1, l_2, \ldots, l_n$$

intersect. Let l denote the common part of these lines (which is obviously either a point or a line segment). Let us prove that the segment l_{n+1} intersects l. Suppose that this is not the case; then there exists a point A between l_{n+1} and l (Fig. 64a). But each of the

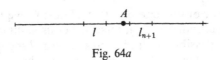

Fig. 64a

line segments

$$l_1, l_2, \ldots, l_n$$

contains l and by agreement intersects the segment l_{n+1}; consequently, each of these line segments contains the point A. Therefore, A belongs to l. The resulting contradiction proves that l_{n+1} and l intersect; their common part belongs to all of the given segments

$$l_1, l_2, \ldots, l_{n+1}.$$

B. In a plane let n circular discs be given such that every three of them intersect. Prove that there exists at least one point common to all of these discs.

Proof. 1. For $n = 3$ the assertion is obvious.

2. Let us assume that our assertion has been proved for n arbitrary circular discs, and suppose that we are given $n + 1$ circular

discs C_1, C_2, ..., C_n, C_{n+1} in the plane. By the inductive hypothesis the n discs C_1, C_2, ..., C_n intersect. Let us denote their common part by C (Fig. 64b) (the "circular polygon" C may consist of

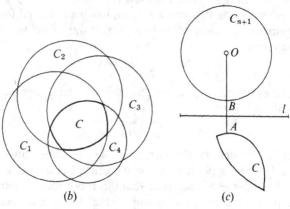

(b)   (c)

Fig. 64

an entire disc or of just one point). We must prove that the figure C and the circular disc C_{n+1} intersect. Let us assume that they do not; then it is possible to draw a straight line l separating the figures C and C_{n+1}. One such straight line would be, for example, the line l perpendicular to the straight line connecting the center O of the disc C_{n+1} with the point A of the figure C nearest to O, drawn at the mid-point of the segment AB, where B is the point of intersection of the segment OA with the circumference of the circle C_{n+1} (Fig. 64c).[1]

Since each of the discs C_1, C_2, ..., C_n contains the figure C and by agreement each intersects the disc C_{n+1}, it follows that each intersects the straight line l as well. Let a_1 denote the segment in which the disc C_1 intersects the straight line l, a_2 the segment in which the disc C_2 intersects this straight line, etc. Then on the straight

---

[1] Indeed, if the straight line l does not separate the figures C_{n+1} and C, then there is a point K on it which belongs to figure C. In the triangle OAK, the angle OAK is acute; moreover by the definition of the point A, $OA = OK$. Consequently, the point L which is the foot of the perpendicular dropped from O to the straight line AK will lie between the points A and K. Since both points A and K belong to all of the circular discs C_1, C_2, ..., C_n, then the entire segment AK does also. Hence, the point L on it also belongs to each of the discs C_1, C_2, ..., C_n, and therefore L belongs to C. It is therefore necessary that $OL \geq OA$. The resulting contradiction ("a perpendicular greater than or equal to a lateral side") proves our assertion.

line l we have n segments $a_1, a_2, \ldots, a_n$. Any two of these segments intersect. Indeed, let us consider two of them, for example, a_1 and a_2. Let M be an arbitrary point of figure C (then the point M belongs to both discs C_1 and C_2). Further, since any three of the given discs intersect, there exists a point N which belongs simultaneously to C_1, C_2, and C_{n+1}. Then the segment MN belongs entirely to the discs C_1 and C_2, and this means that its point of intersection with the straight line l will be a point common to the segments a_1 and a_2.

In accordance with the proposition established in A, on the straight line l there exists a point which belongs to all of the segments $a_1, a_2, \ldots, a_n$. That point must belong to all the discs C_1, $C_2, \ldots, C_n$, and therefore to the figure C, which is contrary to the construction of the straight line l. Consequently, the figures C_{n+1} and C must have at least one common point, which will be a common point for all of the circular discs $C_1, C_2, \ldots, C_n, C_{n+1}$.

C. The proposition formulated at the beginning of this example.
Proof. 1. For $n = 4$ the assertion is obvious.
2. Let us assume that our assertion has been proved for n arbitrary spheres, and suppose that we are given $n + 1$ spheres

$$\Phi_1, \Phi_2, \ldots, \Phi_n, \Phi_{n+1}.$$

Let Φ denote the intersection of the n spheres

$$\Phi_1, \Phi_2, \ldots, \Phi_n$$

(which exists by virtue of the inductive hypothesis). Then, just as in Example 32B, it can be shown that if the sphere Φ_{n+1} does not intersect Φ, then there exists a plane π which separates Φ_{n+1} and Φ. The figures in which the spheres

$$\Phi_1, \Phi_2, \ldots, \Phi_n$$

intersect the plane π are circles, any three of which intersect one another; consequently, there exists a point on the plane π which belongs to all of these circles and therefore belongs to Φ. But this contradicts the definition of the plane π.

The proposition of Example 32 can also be proved by direct induction on the number of figures, instead of by induction on the number of dimensions.

EXAMPLE 33. Prove the proposition of Example 32B by induction on the number of circular discs.

Proof. We shall prove the corresponding proposition for "circular polygons," that is, for figures each of which is the intersection of a finite number of circular discs; from this, in particular, our initial assertion will follow.

1. For $n = 3$ the assertion is obvious.

Suppose we are given four circular polygons C_1, C_2, C_3, and C_4 such that any three intersect. Let A_1 be a common point of the figures C_2, C_3, and C_4; A_2 a common point of the figures C_1, C_3, and C_4; etc. There are two possible cases.

(*a*) One of the points A_1, A_2, A_3, A_4, for example A_4, belongs to the triangle formed by the other three points (Fig. 65a). Then, since the entire triangle $A_1A_2A_3$ is contained in C_4, the point A_4 also belongs to C_4 and, consequently, A_4 is a point common to all four of the figures C_1, C_2, C_3, C_4.

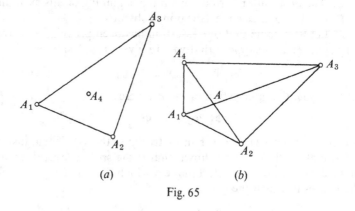

(*a*)   (*b*)

Fig. 65

(*b*) No one of the points A_1, A_2, A_3, A_4 belongs to the triangle formed by the remaining points. In this case the point A, which is the point of intersection of the diagonals of the (convex) quadrilateral $A_1A_2A_3A_4$ (Fig. 65b), being a common point for the triangles $A_1A_2A_3$, $A_1A_2A_4$, $A_1A_3A_4$, and $A_2A_3A_4$ will be a point common to all four of the figures C_1, C_2, C_3, C_4.

2. Suppose that our assertion has been proved for n circular polygons. Let us consider $n + 1$ circular polygons C_1, C_2, ..., C_n,

C_{n+1}. Let C denote the intersection of the figures C_n and C_{n+1} (clearly, C is also a circular polygon). Let us prove that any three of the n figures $C_1, C_2, \ldots, C_{n-1}, C$ intersect. Indeed, if C is not one of the three figures, then they intersect by hypothesis. Let us now consider any three of the figures including C, for example C_1, C_2, and C. Since any three of the four figures C_1, C_2, C_n, C_{n+1} intersect, it follows from part 1 that these four figures have a common point, and this common point will be a common point for the figures C_1, C_2, and C. Since any three of the n figures $C_1, C_2, \ldots, C_{n-1}, C$ intersect, then by the inductive hypothesis all of these figures have a common point, which will be a common point for the $n + 1$ figures $C_1, C_2, \ldots, C_n, C_{n+1}$.

Problem 36.   Prove the proposition of Example 32C by induction on n, the number of spheres.

Hint. Prove the corresponding proposition for "spherical polyhedrons," that is, figures each of which is the intersection of a finite number of spheres. The proof of this proposition is analogous to the proof of the assertion of Example 33.

Problem 37.   Let n points $A_1, A_2, \ldots, A_n$ be given in a plane such that the distance between any pair of them does not exceed 1. Prove that all of these points can be contained in a circle of radius $1/\sqrt{3}$ (Young's theorem).[1]

Hint. Prove first of all that any three of these points can be contained in a circle of radius $1/\sqrt{3}$. Then construct a circle of radius $1/\sqrt{3}$ with each of these points as center, and show that any three of these circles will intersect. The common point of all of these circles, which exists by virtue of the results of Example 32B, will be the center of a circle of radius $1\sqrt{3}$ containing all of the given points.

Problem 38.   Let n points $A_1, A_2, \ldots, A_n$ be given in space, such that the distance between any pair of them does not exceed 1. Prove that all of these points can be contained in a sphere of radius $\sqrt{6}/4$.

Hint. The proof is analogous to the solution of Problem 37.

---

[1] J. Young was an English mathematician of the 19th century.

27. HALF-LINES AND HALF-SPACES

EXAMPLE 34. Let us consider some finite number of half-spaces[1] which fill all of space. Prove that it is possible to select four (or fewer) half-spaces from among them so that these fill all of space.

Let us consider the following sequence of problems:

A. The entire straight line is covered by some finite number of half-lines. Prove that it is possible to select two of the half-lines which will cover the line completely.

Proof. Let A be the vertex farthest to the right of all the half-lines which extend to the left, and B the vertex farthest to the left of all of the half-lines which extend to the right. Since, by assumption, the line is completely covered by the half-lines, the point B cannot lie to the right of A, and the two half-lines with vertices at the points A and B will cover the line completely.

B. The entire plane is covered by some finite number n of half-planes.[2] Prove that it is possible to select two or three of the half-planes which will cover the plane completely.

Proof. We shall carry out the proof by induction on n, the number of half-planes.

1. For $n = 3$ the assertion is obvious.

2. Let us assume that our assertion is valid for n half-planes, and suppose that we are given $n + 1$ half-planes $F_1, F_2, \ldots, F_n, F_{n+1}$ which cover the plane completely. Let us denote the boundaries of these half-planes by $l_1, l_2, \ldots, l_n, l_{n+1}$, respectively. Two cases are possible.

Case 1. The straight line l_{n+1} is completely contained in one of the given half-planes, say in F_n. Then the straight lines l_n and l_{n+1} are parallel. If the half-planes F_n and F_{n+1} lie on different sides of their boundaries (Fig. 66a), then the two half-planes F_n and F_{n+1} cover the plane completely.

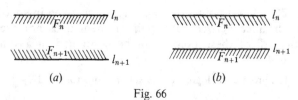

(a)   (b)

Fig. 66

[1] That part of space lying on one side of a plane is called a *half-space*.
[2] That part of the plane lying on one side of a straight line is called a *half-plane*.

In the contrary case one of these two half-planes is completely contained in the other (for example, F_{n+1} is contained in F_n; Fig. 66b), and the theorem follows from the inductive hypothesis, since in this case n of the half-planes (in our case $F_1, F_2, \ldots, F_n$) cover the plane completely.

Case 2. The straight line l_{n+1} is not contained in any one of the half-planes $F_1, F_2, \ldots, F_n$. Then this line is completely covered by these half-planes and they cut off on it $m \leq n$ half-lines which cover the line completely. As we have seen in part A, we can select two of these half-lines which will also cover the line completely. Let F_{n-1} and F_n be the corresponding half-planes. Let us now examine separately two possible cases of the mutual arrangement of the half-planes F_{n-1}, F_n, and F_{n+1}.

(a) The half-plane F_{n+1} contains the point of intersection of the straight lines l_{n-1} and l_n (Fig. 67a). In this case the three half-planes F_{n-1}, F_n, and F_{n+1} cover the plane completely.

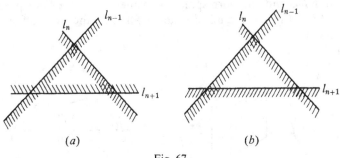

(a) (b)

Fig. 67

(b) The half-plane F_{n+1} does not contain the points of intersection of the straight lines l_{n-1} and l_n (Fig. 67b). In this case the plane is covered by the n half-planes $F_1, F_2, \ldots, F_n$, and the theorem follows from the inductive hypothesis.

C. The original proposition.

Proof. The proof will be carried out by induction on n, the number of given half-spaces.

1. For $n = 4$ the assertion is obvious.

2. Let us assume that our assertion is valid for n half-spaces, and suppose that we are given $n + 1$ half-spaces $V_1, V_2, \ldots, V_n$, V_{n+1}. Let us denote the boundaries of these half-spaces by $\pi_1, \pi_2, \ldots, \pi_n, \pi_{n+1}$, respectively. Then two cases are possible.

Case 1. The plane π_{n+1} is completely contained in one of the half-spaces $V_1, V_2, \ldots, V_n$, for example, V_n. In this case the planes π_{n+1} and π_n are parallel. If the half-spaces V_{n+1} and V_n lie on opposite sides of their boundaries, then these two half-spaces completely fill all of space. In the contrary case, one of the two half-spaces V_{n+1} and V_n is completely contained in the other, and the theorem follows from the inductive hypothesis.

Case 2. The plane π_{n+1} is not contained in any one of the half-spaces $V_1, V_2, \ldots, V_n$. Then it is completely covered by these half-spaces which intersect it in $m \leq n$ half-planes $F_1, F_2, \ldots, F_m$. From the result of part B it is possible to select from among these half-planes two or three which also cover the plane completely (Fig. 66a and Fig. 67a).

Let us examine separately each of the possible cases which may result.

(a) The plane π_{n+1} is covered by two half-planes (Fig. 66a), say F_1 and F_2, and the corresponding planes π_1 and π_2 are parallel (Fig. 68a). In this case all of space is filled by the two half-spaces V_1 and V_2.

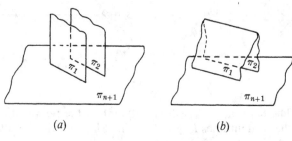

(a)   (b)

Fig. 68

(b) The plane π_{n+1} is covered by the two half-planes F_1 and F_2, but the corresponding planes π_1 and π_2 intersect (Fig. 68b). If the half-space V_{n+1} contains the line of intersection of the planes π_1 and π_2, then the three half-spaces V_1, V_2, and V_{n+1} fill the space completely. In the contrary case the half-space V_{n+1} is covered by the half-spaces V_1 and V_2, and the theorem follows from the inductive hypothesis.

(c) The plane π_{n+1} is covered by three half-planes (Fig. 67a), say F_1, F_2, and F_3, and the plane π_3 is parallel to the line of intersection of the planes π_1 and π_2 (the corresponding planes form a "prism"; see Fig. 68c). In this case the three half-spaces V_1, V_2, and V_3 fill the space completely.

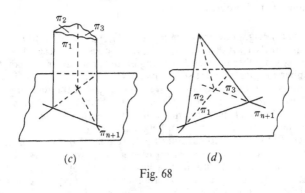

(c)   (d)

Fig. 68

(d) The plane π_{n+1} is covered by the three half-planes F_1, F_2, and F_3, and the plane π_3 is not parallel to the line of intersection of π_1 and π_2 (the corresponding planes form a "pyramid"; see Fig. 68d). If the half-space V_{n+1} contains the point of intersection of the planes F_1, F_2, and F_3, then the four half-spaces V_1, V_2, V_3, and V_{n+1} fill the space completely; in the contrary case the half-space V_{n+1} is covered by the half-spaces V_1, V_2, and V_3, and the theorem follows from the induction hypothesis.

Problem 39. Prove that in space there cannot exist more than four half-lines with the property that the angle determined by any pair is obtuse.

Hint. Suppose that we are given in space some finite system of half-lines such that for each pair, the angle they determine is obtuse. Assume that this system is maximal, in the sense that there does not exist another half-line which makes an obtuse angle with each of the given half-lines. Along with each half-line let us consider the half-space bounded by a plane perpendicular to this at its end point and lying on the same side of the plane as the half-line. Since the half-line system is maximal, these half-spaces will fill the space completely, and our assertion follows from the result of Example 34.

28. ENCLOSING POLYGONS IN SPHERES

EXAMPLE 35.   Prove that there exists a number C_3 such that the sides of any polygon in space $A_1 A_2 \ldots A_n$, none of which are of length greater than 1, can be rearranged (without changing their magnitude or their direction) in such a way that the new polygon thus obtained may be contained in a sphere of radius C_3.

As before, we consider first the corresponding "one-dimensional" and "two-dimensional" problems.

A.   On a straight line are given n points $A_1, A_2, \ldots, A_n$, and the length of each segment $A_1 A_2, A_2 A_3, \ldots, A_{n-1} A_n, A_n A_1$ does not exceed 1. Prove that there exists a number C_1 (not depending on the positions of the points or on the number n), such that the segments $A_1 A_2, A_2 A_3, \ldots, A_{n-1} A_n, A_n A_1$ may be rearranged on the straight line in such a way that the resulting "broken line" $B_1 B_2 \ldots B_n B_1$, each of whose segments coincides in magnitude and direction with one of the segments of the "broken line" $A_1 A_2 \ldots A_{n-1} A_n A_1$, lies completely inside a segment of length $2C_1$.

Proof. Let us agree to consider the length a_i of the segment $A_i A_{i+1}$ of our "broken line" $A_1 A_2 \ldots A_n A_1$ positive if the point A_{i+1} lies to the right of A_i and negative if the point A_{i+1} lies to the left of A_i ($i = 1, 2, \ldots, n$; where it is understood that A_{n+1} represents the point A_1). (We assume that the straight line on which all these points lie is horizontal.) Clearly $a_1 + a_2$ is the length of the segment $A_1 A_3$ (which, in accordance with our agreement, may be either positive or negative); $a_1 + a_2 + a_3$ is the length of the segment $A_1 A_4$, etc.; $a_1 + a_2 + \cdots + a_{n-1}$ is the length of the segment $A_1 A_{n-1}$, and $a_1 + a_2 + \cdots + a_{n-1} + a_n = 0$ (this is the length of the "segment" $A_1 A_1$). Since each segment of the broken line $B_1 B_2 \ldots B_n B_1$ is equal to some segment of the original broken line $A_1 A_2 \ldots A_n A_1$, our proposition may be formulated as follows:

Given n positive and negative numbers $a_1, a_2, \ldots, a_{n-1}, a_n$ such that the absolute value of each does not exceed 1 and such that their sum is 0, prove that these numbers can be arranged in some order $a_{i_1}, a_{i_2}, \ldots, a_{i_{n-1}}, a_{i_n}$ ($i_1, i_2, \ldots, i_{n-1}, i_n$ being a permutation of the numbers $1, 2, \ldots, n-1, n$) such that each of the sums $a_{i_1}, a_{i_1} + a_{i_2}, a_{i_1} + a_{i_2} + a_{i_3}, \ldots, a_{i_1} + a_{i_2} + \cdots + a_{i_{n-1}}$ does not exceed some number C_1 in absolute value (where C_1 does not depend on the sequence $a_1, a_2, \ldots, a_n$ nor even on the value of n).

Let us prove that C_1 may be taken equal to 1. Let $a_1', a_2', \ldots, a_p'$ be all of the positive numbers of our sequence $a_1, a_2, \ldots, a_n$ and $a_1'', a_2'', \ldots, a_q''$, all of the negative numbers ($p + q = n$). Let us select only as many of the first positive numbers $a_1', a_2', \ldots, a_k'$ ($k \leq p$) as will add up to not more than 1 (for example, only the first number a_1'); then add only as many of the negative numbers $a_1'', a_2'', \ldots, a_l''$ ($l \leq q$) as will make the sum of all of the numbers chosen negative, but in absolute value not greater than 1. Then we again add positive numbers and so on alternately until all of the given numbers are exhausted. The sequence obtained in this way, $a_1^* = a_1', a_2^* = a_2', \ldots, a_k^* = a_k'; a_{k+1}^* = a_1'', a_{k+2}^* = a_2'', \ldots, a_{k+l}^* = a_l'' \ldots$, possesses the desired property.

B. In the plane let the polygon $A_1 A_2 \ldots A_n$ be given (it may be non-convex or even self-intersecting) such that the length of any of its sides does not exceed 1 (Fig. 69a). Prove that there exists a number C_2 (not depending

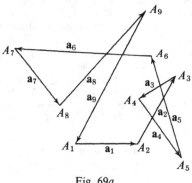

Fig. 69a

on the polygon) such that the sides of the polygon can be re-arranged, not changing their magnitude or direction, so that the resulting polygon $B_1 B_2 \ldots B_n$ lies completely inside a circle of radius C_2.

Proof. Let us prove that C_2 may be taken equal to $\sqrt{5}$. Let the sides of the polygon $A_1 A_2 \ldots A_n$ be considered as directed line segments, that is, as vectors, and denote these vectors by $\mathbf{a}_1, \mathbf{a}_2, \ldots, \mathbf{a}_n$. Let us select some of these vectors, $\mathbf{a}_1', \mathbf{a}_2', \ldots, \mathbf{a}_s'$ so that their vector sum $B_1 B_{s+1}$ (Fig. 69b) has the greatest possible length. Then the projection of each of the vectors $\mathbf{a}_1', \mathbf{a}_2', \ldots, \mathbf{a}_s'$ on the sum $B_1 B_{s+1}$ will have a direction which coincides with the direction $B_1 B_{s+1}$ (for if any one of the vectors had a projection in the opposite direction, then by omitting this vector we could increase the length of the sum).

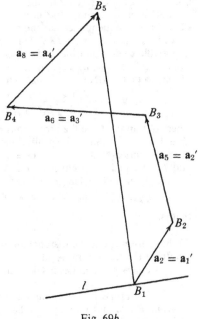

Fig. 69b

Conversely, the projections of all of the remaining vectors, $\mathbf{a}_1''$, $\mathbf{a}_2''$, ..., $\mathbf{a}_{n-s}''$, on $\overline{B_1 B_{s+1}}$ will have the opposite direction. In other words, the projections α_1', α_2', ..., α_s' of the vectors $\mathbf{a}_1'$, $\mathbf{a}_2'$, ..., $\mathbf{a}_s'$ in the direction $\overline{A_1 A_{s+1}}$ may be considered positive $(0 \leq \alpha_i' \leq 1, i = 1, 2, ..., s)$, and the projections α_1'', α_2'', ..., α_{n-s}'' of the vectors $\mathbf{a}_1''$, $\mathbf{a}_2''$, ..., $\mathbf{a}_{n-s}''$ in the same direction negative $(-1 \leq \alpha_j'' \leq 0, j = 1, 2, ..., n - s)$. Further, let β_1, β_2, ..., β_n denote the (positive or negative) projections of the vectors $\mathbf{a}_1$, $\mathbf{a}_2$, ..., $\mathbf{a}_n$ on the straight line l, a line perpendicular to $\overline{B_1 B_{s+1}}$, on which a "positive" direction has been selected. It is clear that $\beta_1' + \beta_2' + \cdots + \beta_s' = 0$ and $\beta_1'' + \beta_2'' + \cdots + \beta_{n-s}'' = 0$ (here β_1' is the projection of the vector $\mathbf{a}_1'$, β_1'' the projection of the vector $\mathbf{a}_1''$, etc.); further, $|\beta_1'| \leq 1$, $|\beta_2'| \leq 1$, ..., $|\beta_s'| \leq 1$ and $|\beta_1''| \leq 1$, $|\beta_2''| \leq 1$, ..., $|\beta_{n-s}''| \leq 1$. By virtue of the result of part A, it is possible to rearrange the numbers β_1', β_2', ..., β_s' and the numbers β_1'', β_2'', ..., β_{n-s}'' so that for arbitrary ν the following inequalities hold true:

$$-1 \leq \beta_1' + \beta_2' + \cdots + \beta_\nu' \leq 1,$$

and

$$-1 \leq \beta_1'' + \beta_2'' + \cdots + \beta_\nu'' \leq 1.$$

If now all of the numbers β_1', β_2', ..., β_s' and β_1'', β_2'', ..., β_{n-s}'' are gathered together in one sequence in such a way that the ordering of the numbers β_1', β_2', ..., β_s' and the ordering of the numbers β_1'', β_2'', ..., β_{n-s}'' is not disturbed, then for any k the sum of the first k of these terms in the new sequence will not exceed 2 in absolute value.

Let us now arrange the numbers α_1', α_2', ..., α_s' and α_1'', α_2'', ..., α_{n-s}'' in one sequence so that the sum of the first k of these terms does not exceed 1 in absolute value. As we have seen from the argument in part A, it is possible to do this *without disturbing the ordering of the subsequences* α_1', α_2', ..., α_s' and α_1'', α_2'', ..., α_{n-s}''. Suppose the corresponding arrangement of our vectors is a_1^*, a_2^*, ..., a_n^*. Then for arbitrary ν we have

$$-1 \leq \alpha_1^* + \alpha_2^* + \cdots + \alpha_\nu^* \leq 1,$$
$$-2 \leq \beta_1^* + \beta_2^* + \cdots + \beta_\nu^* \leq 2,$$

where α_k^* and β_k^* are the projections of a_k^* on $\overline{B_1 B_{s+1}}$ and on l, respectively $(k = 1, 2, ..., n)$. Since the projections of the broken line formed by the sum of the vectors a_1^*, a_2^*, ..., a_ν^* $(\nu = 1, 2, ..., n)$ on the straight lines $\overline{B_1 B_{s+n}}$ and l are equal, respectively, to $\alpha_1^* + \alpha_2^* + \cdots + \alpha_\nu^*$ and $\beta_1^* + \beta_2^* + \cdots + \beta_\nu^*$, then if the length of this vector sum is equal to c_ν,

$$c_\nu^2 = (\alpha_1^* + \alpha_2^* + \cdots + \alpha_\nu^*)^2 + (\beta_1^* + \beta_2^* + \cdots + \beta_\nu^*)^2,$$

that is,

$$c_\nu^2 \leq 5, \quad c_\nu \leq \sqrt{5}.$$

We have found that the distance of an arbitrary vertex of the new broken line from the fixed point A_1 does not exceed $\sqrt{5}$; from this it follows that the entire new polygon lies inside a circle of radius $\sqrt{5}$.

C. The proposition formulated at the beginning of this example.

Hint. Prove that C_3 may be taken as $\sqrt{21}$. The reasoning in this case is analogous to that carried out in part B; β_1, β_2, ..., β_n will be projections of the vectors a_1, a_2, ..., a_n on the plane π, perpendicular to the vector $\overline{B_1 B_{s+1}}$.

29. THE DETERMINATION OF LOCI

EXAMPLE 36. Find the locus of the points in space such that the sum of the squares of their distances from n given points A_1, A_2, ..., A_n is a constant (equal to d^2).

Let us consider the following sequence of problems:

A. On a straight line n points A_1, A_2, ..., A_n are given. Find a point M on the straight line such that

$$MA_1{}^2 + MA_2{}^2 + \cdots + MA_n{}^2 = d^2,$$

where d is a given number.

SOLUTION. Let us take our straight line to be a coordinate axis; let the points A_1, A_2, ..., A_n have coordinates a_1, a_2, ..., a_n, respectively, and let the desired point M have the coordinate x. Then the directed lengths of the segments MA_1, MA_2, ..., MA_n will be equal to $(x - a_1)$, $(x - a_2)$, ..., $(x - a_n)$;

$$MA_1{}^2 + MA_2{}^2 + \cdots + MA_n{}^2 = (x - a_1)^2 + (x - a_2)^2 + \cdots + (x - a_n)^2.$$

$$
\begin{aligned}
(x - a_1)^2 &+ (x - a_2)^2 + \cdots + (x - a_n)^2 \\
&= x^2 - 2a_1x + a_1{}^2 + x^2 - 2a_2x + a_2{}^2 + \cdots + x^2 - 2a_nx + a_n{}^2 \\
&= nx^2 - 2(a_1 + a_2 + \cdots + a_n)x + (a_1{}^2 + a_2{}^2 + \cdots + a_n{}^2) \\
&= n\left(x - \frac{a_1 + a_2 + \cdots + a_n}{n}\right)^2 + (a_1{}^2 + a_2{}^2 + \cdots + a_n{}^2) \\
&\qquad - \frac{(a_1 + a_2 + \cdots + a_n)^2}{n},
\end{aligned}
$$

or, if we let A be the point whose coordinate is $(a_1 + a_2 + \cdots + a_n)/n$,

$$
\begin{aligned}
MA_1{}^2 + MA_2{}^2 &+ \cdots + MA_n{}^2 \\
&= n \cdot MA^2 + (a_1{}^2 + a_2{}^2 + \cdots + a_n{}^2) \\
&\quad - \frac{(a_1 + a_2 + \cdots + a_n)^2}{n}.
\end{aligned}
\tag{15}
$$

$$n \cdot MA^2 = d^2 - (a_1{}^2 + a_2{}^2 + \cdots + a_n{}^2) + \frac{(a_1 + a_2 + \cdots + a_n)^2}{n},$$

$$MA = \pm\sqrt{\frac{1}{n}\left[d^2 - (a_1{}^2 + a_2{}^2 + \cdots + a_n{}^2) + \frac{(a_1 + a_2 + \cdots + a_n)^2}{n}\right]}.$$

Generally speaking, this equality (if the expression under the radical is positive) determines two points M satisfying the condition of the problem (one on each side of the point A).

B. Find the locus of points in the plane such that the sum of the squares of their distances from n given points $A_1, A_2, \ldots, A_n$ is a constant (equal to d^2).

SOLUTION.[1]   Let us select in the plane any rectangular coordinate system, and denote the projections of the points $A_1, A_2, \ldots, A_n$ on the x- and y-axes, respectively, by $A_1', A_2', \ldots, A_n'$ and $A_1'', A_2'', \ldots, A_n''$; let M' and M'' denote the projections of the point M on the coordinate axes. In that case

$$MA_1^2 = M'A_1'^2 + M''A_1''^2,$$
$$MA_2^2 = M'A_2'^2 + M''A_2''^2,$$
$$\cdots\cdots\cdots\cdots\cdots\cdots\cdots$$
$$MA_n^2 = M'A_n'^2 + M''A_n''^2,$$

and, consequently,

$$MA_1^2 + MA_2^2 + \cdots + MA_n^2$$
$$= (M'A_1'^2 + M'A_2'^2 + \cdots + M'A_n'^2)$$
$$+ (M''A_1''^2 + M''A_2''^2 + \cdots + M''A_n''^2).$$

But by virtue of formula (15),

$$M'A_1'^2 + M'A_2'^2 + \cdots + M'A_n'^2$$
$$= n \cdot M'A'^2 + (a_1^2 + a_2^2 + \cdots + a_n^2)$$
$$- \frac{(a_1 + a_2 + \cdots + a_n)^2}{n},$$

$$M''A_1''^2 + M''A_2''^2 + \cdots + M''A_n''^2$$
$$= n \cdot M''A''^2 + (b_1^2 + b_2^2 + \cdots + b_n^2)$$
$$- \frac{(b_1 + b_2 + \cdots + b_n)^2}{n},$$

where $a_1, a_2, \ldots, a_n$ and $b_1, b_2, \ldots, b_n$ are the abscissas and the ordinates of the points $A_1, A_2, \ldots, A_n$, and A' and A'' are the

---

[1] A different solution for this problem has been outlined earlier (see Problem 20).

points on the x- and y-axes with the coordinates

$$(a_1 + a_2 + \cdots + a_n)/n \quad \text{and} \quad (b_1 + b_2 + \cdots + b_n)/n,$$

respectively. Thus,

$$
\left.
\begin{aligned}
MA_1{}^2 &+ MA_2{}^2 + \cdots + MA_n{}^2 \\
&= n \cdot MA^2 + (a_1{}^2 + a_2{}^2 + \cdots + a_n{}^2) \\
&\quad + (b_1{}^2 + b_2{}^2 + \cdots + b_n{}^2) \\
&\quad - \frac{(a_1 + a_2 + \cdots + a_n)^2}{n} \\
&\quad - \frac{(b_1 + b_2 + \cdots + b_n)^2}{n},
\end{aligned}
\right\} \quad (16)
$$

where A is the point of the plane whose projections on the coordinate axes are A' and A'', respectively. From this it follows that

$$
\left.
\begin{aligned}
n \cdot MA^2 &= d^2 - (a_1{}^2 + a_2{}^2 + \cdots + a_n{}^2) \\
&\quad - (b_1{}^2 + b_2{}^2 + \cdots + b_n{}^2) \\
&\quad + \frac{(a_1 + a_2 + \cdots + a_n)^2}{n} \\
&\quad + \frac{(b_1 + b_2 + \cdots + b_n)^2}{n},
\end{aligned}
\right\} \quad (17)
$$

and, therefore,

$$MA = \sqrt{\frac{Q}{n}},$$

where Q denotes the expression on the right side of equation (17). That is, the locus we are seeking is a circle of radius $\sqrt{Q/n}$ if $Q > 0$; it is the single point A if $Q = 0$; and it has no points if $Q < 0$.

C. The proposition formulated at the beginning of this example.

Hint. Consider a three-dimensional cartesian coordinate system with x-, y-, and z-axes, and project all of the points onto the xy-plane and also onto the z-axis; then use formulas (15) and (16).

30. DEFINITIONS BY INDUCTION ON THE NUMBER OF DIMENSIONS

EXAMPLE 37. Define the medians and the center of gravity of a tetrahedron.

A. The *center of gravity of a segment* is its mid-point.

B. A *median of a triangle* is a line segment joining any one of its vertices with the center of gravity of the opposite side. It is well known that the medians of a triangle intersect at a common point; this point is called the *center of gravity of the triangle*.

C. A *median of a tetrahedron* is a line segment joining any one of its vertices with the center of gravity of the opposite face.

Let us prove that the medians of a tetrahedron intersect at a common point. We consider the tetrahedron $ABCD$ (Fig. 70), and let O_1, O_2, O_3, and O_4 be the centers of gravity of the triangles DBC, ACD, ABD, and ABC, respectively. Since the straight lines BO_1 and AO_2 intersect at P, the mid-point of the segment CD, then the straight lines AO_1 and BO_2 intersect at some point O_{12}. Analogously, AO_1 and CO_3, AO_1 and DO_4, BO_2 and CO_3, BO_2 and DO_4, CO_3 and DO_4 intersect at points O_{13}, O_{14}, O_{23},

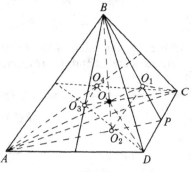

Fig. 70

O_{24}, and O_{34}, respectively. Let us prove that all of these points coincide (in the figure at the point O). Suppose, for example, that O_{12} and O_{13} did not coincide; then the straight lines AO_1, BO_2, and CO_3 would all lie in a single plane π (in the plane $O_{12}O_{13}O_{23}$), but then the straight line DO_4, intersecting AO_1, BO_2, and CO_3, would also lie in the same plane; that is, all four vertices of the tetrahedron would have to lie in the one plane π. Since this is not the case, the points O_{12} and O_{13} must coincide, and in the same way all the remaining points O_{14}, O_{23}, O_{24}, and O_{34} also coincide with this point.

The point of intersection of the medians of a tetrahedron is called the *center of gravity of the tetrahedron*.

Problem 40. Prove that the center of gravity divides every median of a tetrahedron in the ratio $3:1$ with the longer portion toward the vertex.

Hint. Use the facts for a triangle (see Example 26).